A VIVID SHAFT OF NORTHERN LIGHT

Journeys with
Bishop Johan Ernst Gunnerus and
Flora Norvegica through three centuries

Gertrude M. Marsh

Gertrude M. Marsh

A VIVID SHAFT OF NORTHERN LIGHT

*Journeys with Bishop Johan Ernst Gunnerus and
Flora Norvegica through three centuries*

Academic Press

© Tapir Academic Press, Trondheim 2002

ISBN 82-519-1777-8

This publication may not be reproduced, stored in a retrieval system or transmitted in any form or by any means; electronic, electrostatic, magnetic tape, mechanical, photo-copying, recording or otherwise, without permission.

Layout: Tapir Academic Press
Printed by PDC Tangen AS

Coverphoto: Ivar Mølsknes

Tapir Academic Press
N–7005 TRONDHEIM

Tel.: + 47 73 59 32 10
Fax: + 47 73 59 32 04
E-mail: forlag@tapir.no
http://www.tapir.no/forlag

CONTENTS

Dedication 7

Preface Bishop Finn Wagle 9

Acknowledgements 11

Introduction 13

Chapter 1 15
    Christiania, Norway 1718–1740

Chapter 2 27
    Copenhagen, Denmark (I) 1740–1758

Chapter 3 43
    Trondheim, Norway (I) 1758–1759

Chapter 4 59
    North Norway 1759–1760

Chapter 5 71
    Trondheim (II) 1760–1763

Chapter 6 81
    Trondheim (III) 1764–1766

Chapter 7 95
    Trondheim (IV) 1767–1771

Chapter 8 103
    Copenhagen (II) 1771–1772

Chapter 9 109
    Trondheim (V) 1772–1773

Chapter 10 117
    Gunnera

Chapter 11 123
    Flora Norvegica From 1766

Chapter 12 129
    Postscript 2002

This Planet F. Pratt Green 139

List of illustrations 141

Bibliography 143

Gertrude Mary Marsh's Scandinavian journeys 1952–2002

DEDICATION

DEDICATED

To the memory of all whose directions, like signposts, have pointed our way to the present, without any recognition;

Who set stepping-stones that we may be lifted above the relentless flow of the stream of time as we press on to the future, without any praise or reward.

Thanks be to God!

Viola odorata L.
Marsh violet
From Flora Danica

PREFACE

Carl von Linné in one of his letters to the Bishop of Nidaros, Johan Ernst Gunnerus, stated, "There never has been and never will be a bishop like you. Only you know both of God's books, the book of relevations and the book of nature.... This is more evident here in the high north than among any other bishops in Europe. May the Holy Trinity keep you fit and healthy".

As things turned out, Linné's final wish was not heeded and Bishop Gunnerus died at the age of 55 in Kristiansund on one of his visitations. By then he had been bishop for 15 years and had a diocese that extended across half of Norway. With his incredible capacity for work and his wide fields of interest he made Trondheim the cultural centre of Norway. He had also built up a reputation in Norway and abroad that was unrivalled by any Norwegian bishop since the Reformation. Consequently, these deeds support Linné's generous assessment. It should also be remembered that Gunnerus, the first person who really embodied the theology of the Enlightenment in Norway, has been termed "one of the most universal spirits that has been bred by the nation".

"Only you know both of God's books" wrote Linné, "the book of relevations and the book of nature". This is the key to understanding Gunnerus' life as a bishop. He faithfully carried out his prescribed duties based on God's book of relevations. However, at the same time, and as he got to know his diocese, he developed into a noteworthy natural scientist with a burning interest in God's other book: nature. This involvement was something that he was able to cultivate and develop during his visitation trips. In fact, during a period of 11 years he took four extended and strenuous visitation trips to northern Norway and only one to part of his diocese in mid-Norway. Does this mean that the running of Christian life required so much extra effort on his part in the north? I doubt it. It was surely his interest in nature that was his main motivation.

This is a fascinating and exciting life that Gertrude M. Marsh has tried to capture in her book. Something she has done with great success. It is a travel book with exceptional vivid quality, as its subtitle reminds us: *Journeys with Bishop Johan Ernst Gunnerus and Flora Norvegica through three centuries*. Her interest, enthusiasm and knowledge have made Gertrude Marsh a faithful guide along the footsteps of Gunnerus through the various phases of his exceptional life and work. The result is a travel book that reveals the very nature that inspired Gunnerus' study of natural science and, at the same time, it is a fascinating door opener to part of Norwegian cultural history.

Perhaps it is no coincidence that this book is published just now. We face an era where a long line of ecological challenges threaten us and the cries from our wounded planet cannot remain unheeded. Gertrude Marsh is more than a guide to Bishop Gunnerus and his world. She is just as inspired and devoted to God's other book: nature, as he was. Consequently, this is a book with an ecological message. From page one onwards we are presented with the value and richness of creation. Nature is not just natural resources. Nature is a gift that has to be nurtured with humility and gratitude. When we grasp this, we have reached the core of Bishop Gunnerus' work and the source of Gertrude Marsh's inspiration.

Bishop of Nidaros, 4th Sunday after Easter 2002

Finn Wagle
in the Bishop's seat of Nidaros

ACKNOWLEDGEMENTS

The author is most grateful to

His Grace, Bishop Finn Wagle of Nidaros, who in the direct line of Gunnerus' successors contributed the Preface.

Editors: Dr Alan Parton (in Warwickshire) and Dr Stein Johansen (in Trondheim) for their infinite patience and wisdom in guiding the prepartion of this book;
Advisers: Monica Aase (Historian), Karin-Helene Hognestad (Theologian); Harald Nissen (The Royal Norwegian Society of Sciences and Letters);
Consultants: Associate Professor Peter Wagner (Copenhagen); Edvin Grådal (Røros).

Helen and Erling Finne, without whose encouragement, patience and insistence this book would have remained a hidden script;

Margaret and Bruce Bain for years of guidance, friendship and knowledge;

Raffles Institution, Singapore - may my friends there live for a few moments in the contrasting arctic in the book's pages;

Warwickshire College, England - the Principal, Governors and particularly the Looi Horticultural library staff, who have catapulted me into my third professional career and abandoned me to modern techniques; to Trevor Lovatt and Paul Hetherington for professional assistance with illustrations;

Clare Wright (plus), most sincere thanks for her continued support in so many ways and at so many times;

Johan Ernst Gunnerus, as represented by the staff of the Norwegian University of Science and Technology Library, particularly Ingar Lomheim, the Director and especially the keepers who meticulously care for the Gunnerus Library and collection.

My friends Mari, Stein and Torbjørn, who have helped in this production, far beyond the call of duty;
And many other friends included in the Dedication.

Alfred Heaton Cooper (AHC)

It is a great privilege to express my personal gratitude to John Heaton Cooper and the *Heaton Cooper Studio*, Grasmere, Cumbria, England. Our joint affinity is again, Norway, for Alfred Heaton Cooper married Mathilde Marie Valentinsen, daughter of Rasmus Valentinsen the wool dyer of Balholm (now Balestrand) on Sognefjord, in 1894. Alfred's son William was a very popular Lakeland artist but on my first visit to the studio in 1956 it was the painting of Norway by Alfred that attracted my attention.

It was only in 1994/95 that I discovered the Norwegian connection and have since built up my personal collection of Alfred Heaton Cooper paintings of Norway. Alfred was born in the Lancashire cotton town of Bolton in 1864. The young artist's first visit to Norway was in 1891 and I am delighted to add reproductions from his sketchbook of that time, only recently discovered amongst the attic clutter, to fill a gap between Gunnerus' journeyings and my own photographs. The Heaton Cooper tradition continues into the third and fourth generations, landscape painters 'par excellence'.

Gertrude M. Marsh
Warwickshire/Singapore/Trondheim 2002

INTRODUCTION

"Du er heilnorsk", (You are a dyed-in-the-wool Norwegian), a Norwegian friend said to me decades ago, paying me an unexpectedly welcome compliment. I had never considered it before though I realised I had felt it subconsciously since childhood. I had always known the meaning of 'dale', 'fell' and 'foss', for some of my earliest steps were taken with a dear grandfather who resided almost thirty years in Wharfedale, Yorkshire. It was a treat for the little girl, brought up in a smoky, industrial town, to take the miniature, oval wicker basket and explore the pinewoods and limestone area for precious primroses, shy snowdrops with drooping heads and rare violets.

From Wharfedale I departed on my first visit to Norway in August 1952, little dreaming it would be the first of many, the commencement of a slowly developing, lasting love-affair. Whilst prevented from visiting Norway for almost twenty years (1975–1994), being resident in Singapore, an

PRIMROSE Primula vulgaris

Introduction

KILNSEY CRAG. Upper Wharfedale, Yorkshire, the massive limestone outcrop which dominates the dale. AHC.

WINTER. Icicles formed from water draining over rocks beside the road up Gauldalen.

An inner fjord (Sunnylven) near GEIRANGER. Late summer.

island at the tip of the Malay Peninsula, my interest in tropical plants evolved. Yet secretly I planned many an imaginary dream journey to Norway with the aid of a map. By this time I had travelled the length and breadth of the country with a camera and could picture the seasonal changes.

Seasonal changes! Unknown in my Singapore environment where there is perpetual high summer and day and night are of equal duration throughout the year. How could you explain the symbols of Christmas to people never experiencing the cold darkness of winter and the hazards of sleet, ice and snow; or the springtime renewal of life to those enveloped in humidity and consequently continual greenery? Apparently the world is smaller than ever today and transportation from zone to zone so fast that at times our bodies rebel. We have the opportunity to see and discover for ourselves the wonders of natural science through films, words or experience. The knowledge of those who have gradually unwrapped the secrets makes us realise how much there is still waiting to be revealed.

With the opportunity to travel again to my beloved terrain and the heightened interest in wild flowers not seen for decades and now becoming rare, I sought a detailed guide. When I enquired about an illustrated Scandinavian Flora someone said, "You need Gunnerus but you won't find one!" Yet I did! After discovering Johan Ernst Gunnerus', 'Flora Norvegica', in an antiquarian bookseller's in Oslo, I was enraptured and now must find Gunnerus himself.

1

Christiania, Norway
1718 - 1740

Early Days

The daily period of grey light was lengthening in February but the bitter cold intensified, forcing a way into timber and stone homes clustering round the old fortress. The sharp northeasterly wind from the high mountainous hinterland attempted to push its unbidden way into every passageway and courtyard and then teasingly release its boisterous energy on the sea.

The town of Christiania nestled at the head of the hundred-mile deep-sea inlet gouged out by ice millennia ago. The fjord provided sheltered protection from the vicious Skagerrak, the swirling shipping way between the northerly township and the dominant city of Copenhagen, the capital of the Kingdom of Denmark-Norway. Throughout the eighteenth century Norway, the northernmost tip and western half of the Scandinavian Peninsula, including Christiania, was ruled from the capital, Copenhagen.

Christiania, Norway 1718–1740

The winter road down the valley.

For centuries scattered settlements had been established round the inner fjord for there was fertile soil for the cultivation of crops and the support of cattle. Fresh water was available from the country's eastern valleys running from the perpetually snow covered mountains in the north to the sea. There was supplemental fishing during part of the year with any excess catch for salting or drying. Yet it was the availability of timber, conveniently at hand and easily transportable along fast flowing rivers to the sea, that was the foundation of trading.

Akershus Fortress, built originally at the commencement of the fourteenth century, to combat marauding Swedes and/or Danes, was strengthened and extended by King Christian IV. Oslo town had been devastated by fire and the King personally directed its' rebuilding in more durable, fire-resistant materials and gave the town his name, Christiania. Christian IV was a great builder! (The name OSLO was only resumed in 1925)

In February 1718, amongst the population of some five thousand souls in Christiania, we find civil servants, soldiers and mainly timber merchants and their servants. During the years following the Great Fire of London in 1666 a thriving export trade in timber had developed along the south coast of Norway, particularly where rivers and fjords joined the sea. Wealthy merchant 'houses' established saw mills. English ships brought salt as ballast and carried back the raw materials for the rebuilding of the gutted city of London.

Included in the professional elite of Christiania at that time was the physician, Erasmus Gunnerus. He had been a military surgeon and had prolonged his foreign service to extend his medical experience. The fact that he was mainly self-educated was concealed by his skill in anatomy gained through military experience and his enthusiasm for scientific matters.

Parentage

On the twenty-sixth day of February 1718, a male child was born in the home of Dr Erasmus Gunnerus and his wife, Anna, to be named Johan Ernst.

Akershus Fortress, Christiania

> (It is interesting to note that some early sources give this date of birth as 16th February 1718 – the ten days' difference being due to the implementation of the Gregorian calendar. Although fixed in 1582 arrangements were not changed until the early eighteenth century, 1744 in Sweden and in Britain September 1752 when eleven days had to be 'lost' to achieve parity. O. Nordgaard, in his address to the Cathedral School, Trondheim on 26th February 1918, the Bicentenary of J.E.Gunnerus' birth, mentioned this discrepancy. I had noted similar referential changes in dates given in the journals of John Wesley in England and Linnaeus in Sweden. I have been surprised, entering a new high-tech millennium more has not been said of the lost 10/11 days in the eighteenth century. The change would not have much effect on a new-born child but it certainly did amongst agrarian populations dependent on solar power.)

Little is known of Dr Gunnerus' family background. He may have been Swedish/Norwegian for during the previous century domination of Norway had alternated between the crowns of Denmark and Sweden and there had been frequent border changes. We know, particularly in Scandinavia, that family names for centuries had consisted of a parent's name and *-son* or *-datter*. By the mid-seventeenth century this format was confusing for scholars entering the few

Timber is still floated down rivers and fjords to old established saw-mills. There has been reforestation in force for more than 200 years in Norway.

Christiania, Norway 1718–1740

Wash day on a nineteenth century seter.
AHC.

universities and so, on registration, they adopted a Latinised form of a family name or area. The form Gunner*us* may suggest his father, or grandfather, had a university education at Lund, Uppsala or Copenhagen – there was no university in Norway.

Fru Gunnerus was Anna Gerhardsdatter (Gerard's daughter) of Scottish descent. Was there timber, wool or fish in her background?

Erasmus Gunnerus would have gleaned some botanical knowledge for he must have collected flowers, herbs, parts of trees and some mineral salts for such formed the basis of a physician's pharmacy in the early eighteenth century. We can visualise the young Johan Ernst accompanying his father to gather material for his pharmacopoeia and gleaning knowledge as they collected, sorted, preserved and dried nature's remedies. It would have been presumed that the eldest son would follow his father's profession.

We can imagine the lively child having great pleasure in assisting his father in his experiments and dissections. Was this where the boy's interest in natural science was stirred? It is not surprising he found such joy in later life in possessing and using a rare microscope, (details later), and dissecting plants for taxonomic identification.

There would be frequent trips into the surrounding countryside for many of the residents of Christiania had farms and estates outside the city boundary. Families would move out from the sheltered warmth and protection of their winter townhouse to oversee the cultivation of crops, – cereals, vegetables, fruit and flax. Many young women and girls would be occupied higher up the mountains attending the cattle and goats and preparing cream, butter and cheese to be stored for the winter and taken down to the farm and into town when required by the family remaining there for their personal use or ships' supplies.

Summertime on a farm above Romsdalsfjord in Western Norway a century ago.
AHC.

Johan Ernst's education was obviously commenced in the home for when he entered the Christiania Cathedral School in 1729, aged 10 years, he made outstanding progress. Jens Worm's Encyclopaedia describes him as "an active prodigy, ahead of his fellow students". He appeared to have an innate ability in Latin and Greek and was soon presenting lectures in Latin to his fellow students, for he found the available textbooks in philosophy inappropriate to the day-and-age.

This promising education was abruptly halted in October 1732 by the sudden death of his father leaving the fourteen-year-old Johan Ernst the mainstay of a large impoverished family.

Youth

The grievous event of the death of Dr Erasmus Gunnerus suddenly left Anna Gerhard Gunnerus and a large family of young children, without any financial support or means of livelihood. There does not seem to be any suggestion of intimate family support for the bereaved family, as is usual at such times or it would have been possible to trace the parental lineage. Most biographers, writing almost two centuries later,

Christiania, Norway 1718–1740

Oslo Cathedral (Christiania).

dismiss father Gunnerus with the trite statement, "Although an esteemed doctor he died in great poverty". This had a deep effect on the education of his family.

Johan Ernst Gunnerus was the eldest of the extensive young family. Mother Anna, it is suggested, had given birth to fourteen children though how many survived is not known. Christiania church records report the baptism of eight children between 9 July 1721 and 12 April 1730 (Johan Ernst was not included in the list). Some Christian names were repeated suggesting earlier infants of that name had died. We do know that at least two sons and two daughters reached adulthood for they appear later in this account.

The most influential personages in Christiania at that time were the Stattholder (Governor, as king's representative) and the Bishop, both appointed by Frederik IV in Copenhagen. Stattholder Ditlev Wibe had been hurriedly sent to Christiania in the spring of 1722. He had been a personal friend of the king until the king married his mistress, three weeks after the death of his wife (Queen Louise), and had her crowned Queen Sofie on May 31, 1721. The resulting disagreement made it necessary for Wibe to be removed immediately from high office in Copenhagen.

On the other hand, Bishop Bartholomaeus Deichman championed the mistress-queen and so remained popular in the royal court. Untroubled by religious scruples he was a wealthy, brilliant administrator, and avid book collector.
The ambitious Deichman persuaded the king to sell Norwegian churches and church property to private individuals

who would be responsible for their upkeep, thus removing the cost from the state yet providing the possibility of a land tax. Six hundred and twenty churches and estates were disposed of at this time. The purchasers considered they had a good deal until the treacherous years following 1725 when many buildings became dilapidated.

The king, fearful of a revolt over land valuation taxes in the north of his kingdom, sent for the bishop and the stattholder, who did not approve of the bishop's policy, to Copenhagen in 1724. Deichman lost his case. In financial straits he had to sell his library.

> (Bishop Deichman's son, Carl, a prominent industrialist, was later able to repurchase some of his father's book collection. On his death he donated the complete library for public use in Christiania.)

In the first decades of the eighteenth century there were still outbreaks of plague on the periphery of Europe brought by ships to the southern ports and occasionally spreading up the valleys. The last serious European outbreak had been in Marseilles in 1720 but remnants could have stretched over a much wider area and have taken a longer period to clear. At this time the Norwegians built their wooden storehouses on pedestals of solid stone so rats could not infest their food supplies with vermin carried disease.

No doubt several infectious diseases would pass as 'plague', though not specifically identified at that time. The physician would be constantly in contact with known and unknown infections and so place his own family at risk. It is recorded

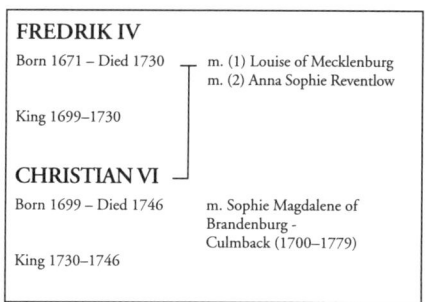

Royal Lineage

Christiania, Norway 1718–1740

in the church records that Erasmus Gunnerus paid for the burial of three 'little children' between 1727 and 1730.

Through the influence of a patron, arrangements were made for Johan Ernst to continue his education at the cathedral school, assisting as a tutor to defray expenses. Some sources name Stattholder Wibe as the patron but he died on 4 October 1731, twelve months prior to the sudden death of Erasmus Gunnerus.

It is likely that Bishop Peter Hersleb of Christiania may have provided significant help in 1732. He was aware of the social conditions and built almshouses for widows and homes for the poor. He did not meddle in politics but sought the welfare of all his flock. In 1736, under the auspices of the pious King Christian VI, he re-introduced 'confirmation' for young people, which had to be preceded by a period of instruction ensuring almost all children could read and write.

> (The result of the bishop's zeal in this matter is still evident in Scandinavia today where confirmation is a prominent event in family life. If you visit any area on a Sunday in May you will find family parties decked in the national costume of the area, travelling by road and ferry to some distant church for the Confirmation of one or two nervously excited teenagers.)

We may assume Bishop Hersleb, during his brief period of influence in Christiania, cared for Widow Gunnerus and her family, for in 1737 he was elevated to the position of Bishop of Sjælland (Copenhagen). During the previous five years he would also have had close contact with the Cathedral School (occasionally referred to as the Latin School) where Johan

The festive dresses, as worn in the county of Rogaland south of Bergen, are hand embroidered with wool on woollen material, using local stuff. Each district has its own style.

Ernst Gunnerus was pursuing his own scholastic studies, fees being covered by his assistance and tutorship within the school.

University faculties of the time were divided between medicine (including a little botany) and philosophy/theology. In the normal course of events this bright young man would have been expected to take up medicine, 'following in father's footsteps' but we find in 1737 Gunnerus has chosen the alternative discipline. The experiences of the five maturing years, following his father's death, had proved his remarkable abilities in teaching philosophy and lecturing. The good Bishop of Christiania had moved to Copenhagen.

Gunnerus elected to study in Copenhagen under the supervision of the Norwegian, Ludvig Holberg, claimed by the Danes as their national playwright. Students entering the university were required to name a tutor to whom they could resort for advice if necessary and many Norwegians chose to enlist with Holberg.

Travelling frequently between Copenhagen and Christiania as a university student, Gunnerus again supported himself by teaching, tutoring and preaching. His powerful voice and simple, direct delivery made him a popular speaker at a time when preachers, particularly in Norway, lacked training and were unsure of the current doctrines.

The Lutheran Reformation had been slow to diffuse through the sparsely populated mountainous regions of the northern territories. The sudden, destructive removal of church treasures, collected over earlier centuries, had not been replaced by

LUDVIG HOLBERG looks out over Bergen, Norway, the city of his birth. He moved to Copenhagen in his early twenties never to return.

Christiania, Norway 1718–1740

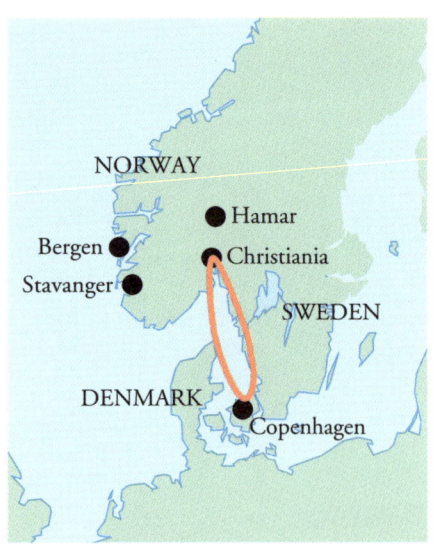

Christiania, Norway to Copenhagen, Denmark.

constructive understanding, especially in isolated communities. What a privilege and pleasure to be able to listen to a young enthusiastic speaker who spoke their language in a manner they could understand!

Enabled to return to Copenhagen in 1740 Gunnerus took the examinations in philosophy and theology. His responsibilities in Christiania had eased as younger members were able to support the family and he was able to remain in the academic world of Copenhagen. But it was a changing world!

Christiania, Norway 1718–1740

Early Eighteenth Century Map of Denmark-Norway and Sweden.

2

Copenhagen, Denmark (I)
1740–1758

Undergraduate

Copenhagen was not an entirely strange place to the twenty-two year old Gunnerus when he became a member of the university there in 1740. He had paid several visits during the previous three preparatory years but now he was to become absorbed in the hub, the very centre of his known world, and the capital of the Kingdom of Denmark.

There was no longer the enveloping arm of mountains in the background but a stretch of flat, wet islands, drained by canals and streams and connected by bridges and ferries. Denmark itself consisted of many, many islands the largest of which, Sjælland, included the city of Copenhagen and the port of Helsingør. The island was well cultivated by several wealthy landowners and hereditary small farmers. Between Jylland (Jutland), the extended flat promontory of northern Europe (including many states such as Hanover, Slesvig and Holstein), was another large island, Fyn. Some of the scattered

Copenhagen, Denmark (I) 1740–1758

Kronborg Castle.
Guardian of the narrow Sound between Denmark and Sweden.

smaller islands were occupied and cultivated but some were just fishermen's havens or sandbanks.

Christiania and the market-ports along the southern coast of Norway had developed in the previous decades, as the timber trade increased. A few foreign ships were usually to be seen and strange stories to be heard from overseas. Here were flotillas of ships, large and small, queuing to pass through the narrow strip of water, the Sound, between the island of Sjælland (Zealand) and Southern Sweden. The narrow entrance to the affluent Baltic Sea gave access to supplies of minerals, hemp and tar, even a route to the fabulous Orient by way of Russia. Sjælland had become rich through the tax levied on foreign vessels passing through the Sound, beneath the mighty canons of Kronborg Castle, the Elsinore of Shakespeare's 'Hamlet'.

Helsingør (Elsinore) may not have been popular with ship owners, because of the tax levied from 1429 to 1857, but it certainly was with the foreign sailors for they knew they were particularly under the care of 'St. Gertrude'. Two altars to St. Gertrude, the special guardian of all travellers and especially sailors had been added to the thirteenth century parish church of St. Olav of Norway. (See later section on Nidaros/Trondheim).

Typical Danish countryside as sketched by AHC, including details of colours to be used in his eventual painting.

The Carmelite monks built a hospital for sailors close to the church and in a niche on the outer wall, facing the Sound, a figure of Saint Gertrude is reputed to have been placed. It was venerated by seamen who drank to St Gertrude the night before any journey in the hope of her especial protection. The wooden statue was originally gilded but the salty atmosphere quickly eroded the colour. It frequently had to be repainted by popular demand.

Gunnerus would frequently pass the historic Kronborg Castle, mix with the foreign sailors and see the ancient church and hospital on his many journeys to and from Norway before encountering the different coloured landscape of the island of Sjælland (Zealand).

> (Danes today still use the expression 'the gilt is off St Gertrude' without knowing its origin, signifying the brightness, or the edge, has been worn off, much as we use the expression, 'the gilt is off the gingerbread'.)

A black and white smartness was apparent in the thatch-roofed farm dwellings dotted near the forests and scattered amongst the brown-green gold cover of the seasonal changing, cultivated land and slightly undulating pastures. Within every dip and hollow, water collected, sometimes forming a lake but always fringed by dense reed beds, providing ample raw material for thatched roofs evident on small and large buildings.

The nearer our traveller approached the city, by sea or land, the more astonishing the sight became. Beautiful rosy-pink brick palaces, several storeys high, were to be glimpsed through beech and oak trees and wrought iron gates.

Brick had been the most convenient building material in this part of the kingdom for centuries. As monks from Germany and France drifted northwards, unable to build monasteries of stone because of the lack of quarries in this moraine covered land, they taught the Danes to make and fire artificial stone from the abundance of clay. The production of roof tiles and floor tiles developed and the manufacturing of the

Jetsmark Church, Jutland with restored murals, one depicting St. Gertrude and the chapel where her 'order' cared for foreign sailors.

Copenhagen, Denmark (I) 1740–1758

Detail of mural showing St. Gertrude and her Chapel..

Detail of early tiles in Helsingør.

beautiful bricks, in various shades and textures evolved, still to be seen eight centuries later.

Local white stone dressings were eventually used to enhance the royal palaces, replacing timber frames and beams. Marble was imported for interior and exterior embellishment of the more lavish buildings. Dutch influence was later evidenced in the carved, curls of decorative gables.

The most awe-inspiring sight, to the young Norwegian, must have been the intricate church-towers decorated with gold leaf and the green copper roofing of impressive buildings still being completed. King Christian IV had rebuilt several monumental buildings on ancient sites of royal homes. Forever renowned as the Great Builder throughout the kingdom of Denmark, he reigned from 1588 to 1648. Gunnerus certainly knew of his architectural prowess in Christiania and Kristiansand, Norway.

A great fire in 1728 had destroyed the university buildings and a considerable area around including the medieval church of Our Lady. So perished within these two edifices many treasured antiques – books, manuscripts, monuments.

Now, safer more solid buildings were replacing them under the auspices of the reigning king, Christian VI.

(Indeed, some of the buildings, opposite the Round Tower, were still in use by the university theology department in 1997. Twisty narrow staircases and connecting corridors on upper floors, above present day shops, connected various sections, library, IT, tutorial rooms and offices.)

The strictly religious Christian VI ascended the throne, on the death of his father (Frederik IV), in 1730. Educated by pietistic German teachers he led a stern, puritanical life in contrast to the slack moral standards of his father. Pietism had grown from a reform movement in the German Lutheran churches during the seventeenth and eighteenth centuries. Its aim was to renew the devotional ideal. The King's attitude was strange to the lively Danes but gradually the reins of Government were tightened and the position of the church enhanced.

Gunnerus must have found it difficult to assimilate the religious/political change then taking place, particularly in Copenhagen. Bishop Hersleb, as previously mentioned, had been appointed to Christiania by the pietistic Christian VI and remained there some five years before being recalled to the King's side as Bishop of Sjælland.

> (John and Charles Wesley, in England, had come under the influence of another development of Luther's church reform from central Europe at the time of Christian VI. They had met some Moravians seeking religious freedom in the New World (Georgia) and later met with the group, similar in ideals to the Pietists. The Moravian meeting in Aldersgate Street, London, on 24th May 1738, proved a life-changing experience for John Wesley, a super-charged catalyst being released in this itinerant evangelist.)

By law, Christian VI enforced his deep religiousness and intellectual pursuits. The theatre, which had blossomed during the previous decades, ceased. He was meticulous and painstaking in his organisation but not particularly thrifty, as evident in the magnificent, French-style furnishing of the ancient Christiansborg Castle, which he rebuilt. (Now used as the present Parliament Building)

The magnificent Fredriksborg Palace rebuilt by Christian IV now holds the national artistic treasures of Denmark (and some of Norway).

Brick dressed with local white stone and copper-roofed.

Copenhagen, Denmark (I) 1740–1758

Seventeenth Century cobbled street in Helsingør with red brick and timber-framed (oak) three storey houses.

Copenhagen University c1752 with the great auditorium on the left. Painting by Rach and Eegberg in National-Museum, Copenhagen.

Trade was booming in both Norway and Denmark. Unfortunately Denmark lacked the natural force of waterpower so readily available in Norway from the mountains down to the sea. In the low-lying land the only sources of energy were still wind and horsepower. It was good pastureland and as the farms and cattle breeding developed there was an increased demand for dairy produce by chandlers and town dwellers. Fishing and shipping were of paramount importance.

It was into this religious and political climate that Johan Ernst Gunnerus came in 1740 to study theology and philosophy. There is no doubt that he had been influenced in his youth by Norwegian bishops for him to change from the traditional profession of his father to the discipline of philosophy. But had the changes in royal attitude been experienced in Norway during the first decade of the reign?

Most university students had to provide food and shelter for themselves by some form of employment, usually in the form of teaching or tutoring. We can presume Gunnerus first contacted his own personal adviser, Ludvig Holberg, for some influence must have been brought to bear on his situation. He resided and was occupied mainly in the home of Count Adam Moltke as tutor to his many children. Influential in royal circles, at this time Moltke had been chamberlain to the Crown Prince for ten years. Adam Gottlob Moltke had been born into a prominent Mecklenburg family in Walkendorf, North Germany, in 1710. The family moved to Copenhagen when he was a child and he was a page to the future King Christian VI.

On the accession to the throne of Christian, Moltke continued in the Royal Household as chamberlain to the young man who was to become Frederik V. A close, lasting friendship developed between them.

Living in close proximity to the pietistic king and court, Gunnerus found it very difficult to assimilate all the changes thrust upon him. Not only were the living conditions entirely different from those he knew in Christiania but also the religious and philosophical ideas enforced were strange to him.

A good scholar, determined to succeed, he secretly made a study of the new regime and eventually became tutor to the children of Thestrup, the professor of philosophy at the University of Copenhagen. A neat and cheap way of getting to the heart of the matter!

After two years of determined study Gunnerus was awarded a Royal Scholarship to study at a German university. On the advice of his tutors he went to Halle. It is interesting to recollect the close ties, especially in the eighteenth century, interwoven in the political and academic histories of Denmark, Northern Germany and England.

From the election of the Elector of Hanover in 1714 as King George I of England, the power of this state spread its organising, legalistic and cultural influences. George II, (Elector George Augustus of Hanover) endowed Göttingen University where teaching commenced in 1734 and inauguration in 1737. The university library became one of the most famous in Europe.

Moltke's Palace in Copenhagen as it stands today. Here Gunnerus was tutor to the Count's many children.

The only 'mountain' in Denmark, Himmelbjerget, 147m, can just be distinguished in the sketch, near Silkeborg, in Jutland.
AHC

Copenhagen, Denmark (I) 1740–1758

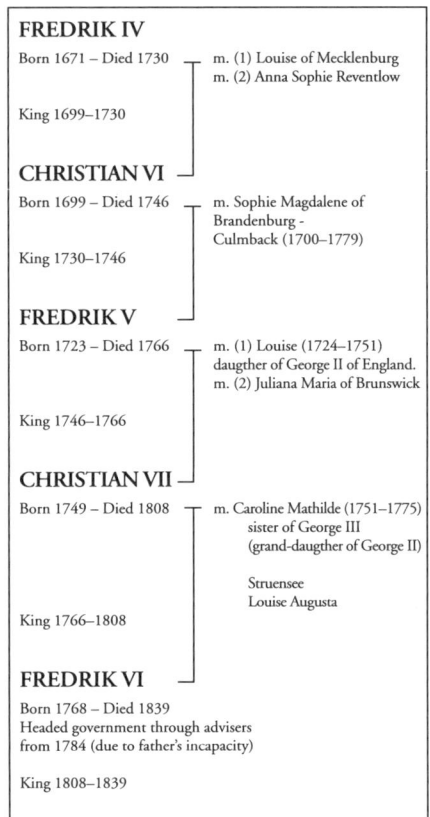

Royal Lineage

An escapee Hanoverian musician wrote the coronation anthem, Zadok the Priest, for George II's coronation and it has been sung at every English coronation since. George Frederick Handel became a naturalized British citizen in 1727 and his oratorio, The Messiah, remains an English 'institution'.

(Not many months ago I was having breakfast in the select breakfast room of the restorative Shangri-La Hotel, Singapore. The soft background music, usually Mozart or Bach, suddenly included a wordless, pianissimo version of the Hallelujah chorus. I stood up, as King George did at the first performance he heard, but nobody else noticed. In December 1997 I had been in the audience both in Copenhagen Cathedral and Huddersfield Town Hall for performances of The Messiah and joined in the customary tribute to the composer. There the similarity ended!)

Several Danish kings had wives descended from the Hanoverian/English royal family at one time or another. Princess Louise, the daughter of George II and Queen Caroline, was the first wife of Frederik V, the cherished friend of Moltke.

Halle

In 1742 Gunnerus travelled south to the cosmopolitan university city of Halle on the River Saale in Saxony. The university, founded in 1694, had become an influential powerhouse of the early period of Enlightenment, attracting faculty and students of international calibre to its precincts. Protestant students gathered there from most parts of northern Europe, from Scandinavia to Russia and southern Germany.

There was no difficulty in supplementing a two and a half year grant by tutoring foreign noblemen's sons completing their education in this academic centre. Not only was Gunnerus' formative classical education and expertise gainfully employed but he was engulfed in the turmoil of strange and conflicting philosophies.

North Germany had been the nursery where the new ideas of the Protestant Reformation had been nourished by the professor of Biblical Theology at Wittenberg. Martin Luther began preaching in 1511 that spiritual forgiveness, a rightness with God, and enlightenment, were not to be bought with money or by deeds, but accepting by faith, and acting upon, the teaching of the New Testament, which was becoming available to all readers.

Christian Wolff, the most influential German philosopher of the time, led the many distinguished writers, teachers and thinkers who congregated at Halle. From 1707 he developed his ideas on mathematics and natural philosophy into what he called, 'the mathematical method' logically based on the two laws of contradiction and sufficient reason.

Wolff was expelled from the university of Halle in 1723, because a lecture he presented on Confucius had antagonised Pietists. He spent the following years writing and publishing his doctrine until he was reinstated in 1740 and at the time Gunnerus was there he was made Chancellor (1743).

SPRING-TIME reflections in Norway. The young leaves are showing on the silver birch tree. (Hardanger, May 01).

Copenhagen, Denmark (I) 1740–1758

1742 Gunnerus travelled south for further study..

Jena

Gunnerus felt that the teaching in Halle had become too narrow-minded and so moved to the even older university of Jena. Students commonly moved from one university to another seeking and expounding knowledge. There is no doubt he had been influenced theoretically in the field of natural science and philosophy but he desired the experimental experience of development in a wider field.

In December 1745 at the early age of twenty-seven, Gunnerus was awarded a Master's Degree in Philosophy at the University of Jena, then, as now, an academic, scientific centre of learning.

He remained in Jena a total of ten years, writing, lecturing, preaching, developing his debating skills and broadening his interests in academic, social and scientific fields. He was offered the position of rector in the Latin School but he had hoped for a situation in the university. On a later occasion he was offered a professorship but this he also declined as now he was looking towards Denmark. He had several works published in Jena on theological, scientific and economical matters and was made a Fellow of Latinske Selskap i Jena (Societas Latina Jenensis).

Suddenly, in 1754, Gunnerus was called to Copenhagen by the prominent minister of state, J.L Holstein, to be appointed, 'extraordinary professor of theology'. He could not leave Jena immediately because the publication of his most recent work was in hand. Not until Easter 1755 was he able to return to Copenhagen, no longer a student but a renowned philosopher, natural scientist, mathematician and a popular lecturer and preacher.

Theology, Copenhagen

A social/political change had swept Denmark during the absence of Gunnerus in Germany. The pious Christian VI had died in 1746 and his son, now King Frederik V, had ascended the throne. Previous strict tensions were released and a violent reaction to the attitudes of the preceding reign prevailed making the pleasure seeking Frederik a popular king.

Matters of state were left in the hands of a select group of very capable, mainly German, noblemen who had been introduced during the previous reign. The king himself was not interested in politics and passed on responsibility, taking credit for the achievement of others, whilst the members of the Danish nobility were busy enjoying themselves again.

The Queen, Louise had been the daughter of the Elector of Hanover who later became King George II of England, died in 1751, leaving a young infant Crown Prince, Christian, barely two years old. Frederik's second wife was Juliana Maria of Brunswick, demonstrating a further strengthening of German ties in Denmark.

The most prominent Foreign Minister was the wise and able Johan Hartvig Ernst Bernstorff whose family estates were found both in Denmark, Hanover and Holstein. He advocated

Count Moltke's country residence at Bregentved.

Copenhagen, Denmark (I) 1740–1758

a policy of peace for Denmark believing it essential to free trade and the development of the state

Frederik's closest friend and intimate adviser remained Count Adam Moltke, in whose home Gunnerus had stayed as a young student a decade before.

Gunnerus was needed in Copenhagen. He could not be appointed professor of theology because there could be only one occupant of a professorial chair at a time. An extra position had to be created for him. Entitled 'extraordinary' professor, his duties were to include those of vicar of Herlufsholm and lecturer in theology and Hebrew at the elite school of the nobility there (Th.Petersen).

Professor Gunnerus' position was looked upon with disfavour in some university circles. There were those who considered he had been given an undeserved rank without having taken the basic examinations they had.

However his years of experience lecturing, writing, teaching and preaching in Jena were eventually accepted as credentials and Johan Ernst Gunnerus was ordained on 18 June 1755 in Copenhagen. He never did take up the situation at Herlufsholm but became provost of a community foundation within the University of Copenhagen where his lecturing and teaching abilities were appreciated.

Within three years Gunnerus had built up a reputation as a popular lecturer in theology, church history, philosophy, mathematics and natural science. He gave as many as four

The chapel wing and hall at Frederiksborg Palace.

lectures a day in these subjects, not only attracting university students but also keenly interested searchers for modern knowledge and thinking. Men came from great distances to hear the young enthusiastic speaker with a clear voice who presented his subject matter in a form they could understand.

Other professors were jealous of his status and critical of his theory. In later decades some of his students became well known in the scientific-religious field. This was the Age of Enlightenment in Denmark. The strict pietism of Christian VI had gradually given way to the more outward looking Frederik V seeking to impress northern Europe. It is never easy for age to make way for youth!

Acting Professor Gunnerus was unexpectedly summoned to an audience with his majesty at the lovely modern palace of Fredensborg a distance from Copenhagen. This was not an ornate, magnificent home like several royal residences but a peaceful retreat as its name suggests, 'Castle of Peace'. (One source gives this date as 30th June 1758)

In nervous anticipation the forty-year-old Gunnerus must have ridden north, through the beech wood forests in full summer leaf, pondering the reason for the command. He knew there were plans being laid for an expedition to study the plants of the Bible, early manuscripts, and the way of life and ancient customs of the unknown Orient, as the lands bordering the eastern coast of the Mediterranean Sea were known at that time. Gunnerus had studied Oriental languages – Greek and Hebrew, and not Chinese or Japanese as the term suggests today.

Fredensborg Palace remains a peaceful retreat for the Royal Family.

A younger fellow countryman, Jens Henrik Strøm, (the younger brother of Gunnerus' friend, Hans Strøm, born in Borgund, Sunnmøre) had been designated the linguist for this expedition the previous year but had then been influenced to withdraw by the jealous, less-able Dane, Von Haven. Was Gunnerus to be drawn into this mission? Surely he was too old to face the unknown rigours of sun-baked Arabia!

> (This was the ill-fated expedition to Arabia Felix, which eventually departed from Copenhagen in 1761 and from which only one of the six participants returned.)

Instead, according to his own account, the king's commission quite over whelmed Gunnerus.

Across the Skagerrak, in the Danish Kingdom of Norway, Dr Frederik Nannestad had been transferred from the most northerly diocese in the kingdom, Trondheim, to the influential bishopric of Christiania. King Frederik V now appointed Johan Ernst Gunnerus to the most ancient see of Trondheim, Bishop Gunnerus of Trondheim.

Once again it would appear that Count Moltke had been influential in this appointment. He could speak from personal experience of Gunnerus' didactic ability and he knew of his interest in natural science. A man of such broad understanding and knowledge would be an asset in the gathering of information from the far north as the search for Denmark's natural resources progressed.

Copenhagen, Denmark (I) 1740–1758

On the thirty first day of July 1758 Johan Ernst Gunnerus was consecrated Bishop of Trondheim, a bishopric which stretched from Romsdal, through Trøndelag to the most northerly tip of Europe and the Russian frontier.

3

Trondheim, Norway (I)
1758–1759

Towards the City

There were two possible approaches to the city of Trondheim, by sea or land. Perhaps the forty year old newly elected bishop wanted to travel the quickest way with time to consider and prepare for his future calling. In July that would be by sea in a comfortable cabin with the long hours of daylight making study and contemplation possible with ease. This voyage could have taken a week and be preceded even by weeks of waiting for a suitable vessel and favourable weather conditions.

There would be a regular horse-drawn coach service from Copenhagen to Christiania necessitating ferries and endless days of journeying along deeply rutted roads without any possibility of exercise or study. The track up the great north-south lush valley of Gudbrandsdalen and over the barren Dovre Fjell, the ancient pilgrim route from Christiania to Trondheim, would be pleasant and provide a foretaste of the country where he would minister.

Trondheim, Norway (I) 1758–1759

The northern ways to Trondheim.

Gunnerus would undoubtedly call in Christiania on his journey north to visit friends and relations. He would wish to take the opportunity to visit the new Bishop of Christiania, recently translated from Trondheim. They were old acquaintances, for a nephew (brorsønn) of bishop Frederik Nannestad, Nicolay Engelhart Nannestad, had exchanged positions with Gunnerus soon after his arrival in Copenhagen from Jena, for their mutual benefit. (Nicolay Engelhart Nannestad later became professor of the High School in Odense.)

Of Johan Ernst Gunnerus' many siblings one sister plays a particular part in his life as bishop. Stine was the widow of Captain Brohier and accompanied by her two daughters, Elise and Agathe, she organised and ran the bishop's home in Trondheim. Her only son, Johan Mathæus Brohier was a naval lieutenant in Copenhagen but died young.

The bishop had never married though one biographer, (*Gunnar Engegård*), intimates there had been an engagement to a beautiful young lady of good family in Copenhagen. After due consideration Gunnerus realised such a relationship was not conducive to his calling and he distanced himself from such arrangement – there is even a suggestion he 'paid her off' with a monetary gift. The whole affair may have been surmise or gossip but no doubt Gunnerus would learn something of the opposite sex from the experience. Sister Stine managed his household affairs and he was foster father to her children.

From the Christiania baptismal register it would seem that Stine was christened 'Catharina Dorothea' on 12 April 1726.

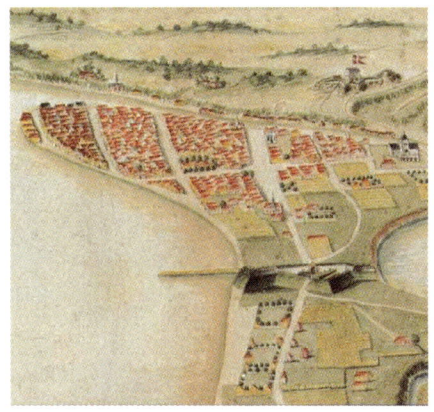

Trondheim c. 1730

As the pronunciation of 'Stine' rhymes with 'Catharina' and 't' and 'h' are separate, we can easily imagine infant voices contracting 'sister Catharina' into the affectionate 'Stine'. There is no record of Johan Ernst's baptism.

The most impressive approach to the ancient city of Trondheim was by sea. Some two hundred and fifty thousand islands, the Skerries, protect the thousand-mile long western shoreline of Norway. No doubt there would be times when the uninhabited rocky outcrops would be dangerous hazards in the eighteenth century without modern navigational aids. But there would be many welcoming havens accessible only by boat where the young bishop would be treated like a visitor from outer space!

After the usually rough stretch of Hustadvika, a vessel would negotiate the wide, island-strewn mouth of the Trondheimsfjord, moving eastwards towards the north-south mountain barrier dividing Sweden from Norway.

The fjord, deep enough for ocean-going vessels, stretches one hundred and thirty kilometres (80 miles) inland before making a great right-angled sweep northward where the fresh mountain waters of the River Nid (*Nidelva*) cross the coastal plain and meander into the sea more or less creating an island.

On this water-protected almost-island on the southern shore of the deeply indented Trondheimsfjord, thirty-seven kilometres southeast of the rough Norwegian Sea, King Olav Tryggvasen (King Olav I) built a church and residence in the

The oldest wooden octagonal church in Norway..

The modern approach - the Express Coastal Steamer departing from Trondheim. Across the fjord can be seen the snow covered mountains.

Trondheim, Norway (I) 1758–1759

Centuries old houses still in use near the harbour in Trondheim.

Fishing craft in Trondheim harbour 2001, old riverside warehouses have been converted into modern apartments.

year 997. The strategic site was an excellent choice being easily accessible by sea-going vessels, the most comfortable and convenient means of transport a millennium ago. Food, drinking water, wool and building materials were readily available in the immediate locality.

A trading post quickly developed, naturally called Kaupanger (*Norwegian kaupang meaning market*) but two decades later it was renamed Nidaros giving prominence to the river, a parent of its birth.

Nidaros gained importance as a pilgrimage centre after legends around the body of King Olav Haraldsson (King Olav II) spread. Olav Haraldsson, later the famous Saint Olav, was buried there after his death at nearby Stikklestad on 29 July 1030. A natural spring evolved at the site and its waters were found to have miraculous curative properties.

The sailors' church in Helsingør (page 29) where the nuns of the Order of Saint Gertrude ministered, was dedicated to this St Olav. Several very similar legends are attributed to both these saints as appearing to storm-tossed sailors and bringing relief in physical distress especially with regard to skin infections.

The first stone church, replacing a wooden structure over St Olav's grave, was built by his nephew, Olav Kyrre (Olav the Quiet), in 1075. Extensive building of the church followed the formation of the bishopric of Nidaros in 1152 and continued into the fourteenth century. The first bishop was an Englishman, Nicholas Breakspear who later became Pope.

With the expansion of the tenth century market of Kaupang-on-Nid, in the Middle Ages the name had gradually developed to *Trondheim*, taking the name of the fjord, but the church remained Nidaros domkirke.

Bishop Johan Ernst Gunnerus arrived in 1758 to a colourful, busy, lively town but a cathedral ravaged over the centuries by fire, storm and tempest and pillaged during the Reformation when gold and silver had been stripped from the church and melted down in Copenhagen.

The diocese for which he was responsible stretched far beyond the comfortable, prosperous city of Trondheim, encompassing Romsdal, the valley of the River Rauma, to the south, including the provinces of north and south Trøndelag to the Russian border and Magerøy, the island tip of unknown Norway to the north. The high mountains of Sweden, the source of powerful waterfalls, extensive lakes and mighty rivers draining into the Gulf of Bothnia formed the often ill-defined eastern boundary. It was sixteen thousand square miles of mainly barren mountains and islands, with narrow coastal lowlands beside coastal indentations and sea-flooded ancient valleys called fjords.

Undaunted, Gunnerus dedicated his interests, experiences and qualifications not only to the academics and social elite of the city but also to the foresters, fishermen and farming communities diversely scattered throughout his see.

The statue of St. Olav, bearing a wreath for Olsok Day (29th July), gazes from the West Front of Nidaros Cathedral (2000).

Trondheim, Norway (I) 1758–1759

JOHAN ERNST GUNNERUS
1718–1773.
Bishop of Trondheim 1758–1773.

In Trondheim

Inducted as the Bishop of Trondheim on 31st July 1758 Gunnerus must have seen a few familiar faces around him. The bishop of Zealand, the diocese including Copenhagen and Roskilde, Dr Ludvig Harboe, previously of Trondheim, performed the Consecration Ceremony.

As Gunnerus had spent almost twenty years in scholarly circles around Copenhagen there must have been some fellow students amongst the Norwegian clergy and schoolmasters he knew. He had the added comfort of his own home attended by his younger sister and her family.

It was certainly very different from Copenhagen. No university! No royal court! Yet it is obvious Gunnerus was soon happily settled accepting his new position though it may have appeared he had been exiled to the most northerly diocese in existence. He took over the previous bishop's house, as no official residence was provided at that time.

Gerhard Schøning

The rector of the Trondheim Cathedral School, who dominated the educational circle, was at that time *Gerhard Schøning*, a dedicated man of the north. Born in 1722 in Skottnes, Buksnes, in the Lofoten Islands off the northwestern coast of Norway and within the Arctic Circle, Schøning himself had studied in the Trondheim Cathedral School, after primary schooling in local Vågan. Inspired by Rector Benjamin Dass' enthusiasm for Norwegian history he went on to Copenhagen University. He studied philosophy and history broadening his knowledge and experience until at the age of twenty-nine he was recalled to Trondheim to succeed Dass on his retirement.

Peter Frederik Suhm

Schøning did not return to Trondheim alone in 1751 but was accompanied by a Danish historian, *Peter Frederik Suhm*, the son of an admiral. Suhm needed financial backing to pursue his academic professional research on the history of Denmark. A helpful friend suggested a good marriage would solve the problem and went so far as to intimate such a suitable young lady was to be found in Trondheim. Within one week of his arrival Suhm was engaged to Karen Angell, the niece of the wealthy Thomas Angell.

Gerhard Schøning

On the surface it would appear that this marriage does not concern Gunnerus. It was not only significant in the deepening of the bishop's philosophy and theological historical interests evolving into natural science but of lasting importance to the city of Trondheim today.

Karen Angell was the only child of Lorentz Angell a direct descendent of Lorentz Mortensen Angell whose family had originally moved from Angeln in Slesvig, a southern province in Denmark, to Norway. Lorentz Mortensen was actually born at Nordgaard, Steinberg Parish, in 1626. He took over farms in Trøndelag in 1650 and at the time of his death, almost half a century later (1695), his extensive properties, interests in mining and timber were shared amongst his four sons.

Peter Frederik Suhm

Trondheim, Norway (I) 1758–1759

Albert, the eldest son, became the principal shareholder in the Røros Copper Mines. His wife, Sarah Hammond, was the only daughter of an Englishman who had married a Norwegian and was well settled in Trondheim. They had two sons, both born within the year 1692, Thomas and Lorentz, but Albert died when they were only twelve (1705, aged 45). Almost twins the two brothers were inseparable. They were educated together at Copenhagen University before further studies and travels in Germany, Holland, England and Ireland.

Returning to Trondheim in 1717 on the death of their mother, who had administered their shared inheritance since the death of their father, the brothers took up their substantial responsibilities. They launched further vast commercial concerns in timber, shipping, importing and exporting and banking. Thomas and Lorentz continued to share their domestic life even after Lorentz married Sara Collett. So Suhm would see Karen as a substantial heiress, a financial saviour.

Lorentz gave permission for the marriage but only on condition Suhm and his wife would remain in Trondheim. He died within weeks of the meeting and so the marriage had to be postponed, for the twelve months' mourning period, until April 1752. The wedding took place in Vår Frue Kirke (Church of Our Lady) and was witnessed by the bride's uncle, Thomas Angell, and the priest Niels Tønder.

When her mother died in 1756, Karen Angell Suhm took over the running of the household, which was now solely supported by Uncle Thomas. This was distasteful to her husband at the time but he made frequent trips to Copenhagen and hoped for

The Sugar-house, one of the oldest factories in Norway, was quite modern when Gunnerus arrived. Sugar cane was imported from the West Indies. It was part of the Angell family business.

future affluence, subsidising his present limited income by writing newspaper and magazine articles. The same year Schøning married the daughter of Jens Hveding, the Town Clerk.

So Gunnerus found on arriving in Trondheim two years later the two learned friends, Schøning and Suhm, meeting at least twice a week in the congenial atmosphere of the Angell family library. Schøning was involved in continuing the work of Dass on the history of Norway. Suhm sustained his writing of Danish history providing a further background study to Schøning's search.

> [Schøning was later referred to as 'the founder of Norwegian Historical Research' and 'a mighty Norwegian Patriot']

Invited to join this select partnership of like minds, the bishop brought his experience of theological philosophy and natural science, his rhetoric and organising abilities and received literary advice and knowledge of the northern territory. It was within this group that the seed was sown for the formation of the Trondheim Scientific Society.

Trondheim Cathedral School c1800 newly rebuilt with money from the Angell Foundation.

Georg Christian Oeder

A visitor in Trøndelag at this time and frequently included in the intellectual group, was the German-born, Danish Royal Professor of Botany, *Georg Christian Oeder*. He would be a welcome associate to Gunnerus for he would not have the permanent relationship with Trondheim Suhm and Schøning held. Gunnerus and Oeder had each received jealously, hostile receptions in Copenhagen academic circles, when the king's

Trondheim, Norway (I) 1758–1759

G. C. Oeder, 1728 - 1791.

Ranunculus lapponicus Oed. From Flora Danica.

advisers had called them from German Universities. The circumstances were resolved by the creation of new positions.

Oeder had been sent to Norway by order of King Frederik V in May 1755, spending almost eighteen months in the area of Christiania. He had proposed the publication of an illustrated flora of all wild plants in the kingdom to enhance the knowledge of their useful and harmful properties. Now his commission was to investigate the natural resources, particularly those of economical value, in the Kingdom of Denmark and Oeder chose to commence in the almost unknown Norway, accompanied by the illustrator, Martin Rössler, the son of the engraver from Nuremberg.

After a period in Copenhagen organising his commitments, Oeder returned in the summer of 1757, spending some time in Christiania and Bergen before moving northwards to Trondheim, making it his base for the exploration of north Norway, just as Gunnerus arrived.

Georg Christian Oeder was born February 1728, ten years after Gunnerus, in Ansbach Bayern, north Germany, where his father, Georg Ludwig, was a schoolmaster and known locally as an evangelical theologian – he is actually listed as such in the Correspondents' List of Carl von Linnaeus. This area was the cradle of the Reformation, for Martin Luther had been professor of biblical theology at near by Wittenberg more than a century before.

At the university of *Göttingen*, established the previous decade, financed by King George II of England, Duke of

Hanover, and structured by Münchhausen, Oeder read medicine and botany, qualifying at the age of twenty-one.

A decisive emphasis was placed on natural science, led by the renowned poet and biologist, Swiss born *Albrecht von Haller*, professor of anatomy, surgery and botany. An exemplary library was built up which became the foremost university library in Germany – it is obvious in Oeder's later life that this collection had made a significant impact.

Dr Oeder commenced his medical practice in Slesvig and within three years the reputation of this lively, sincere, intelligent young man spread as a botanist.

Recommended by Professor Haller, he was called to Copenhagen by J.H.E. Bernstorff, the King's foreign minister, to be professor of botany (1752).

There was student opposition to this appointment, mainly because Oeder was a much younger man than some of his students, one particularly considering himself more suitable for the position (Rottbøll).

There was a murmur of dissent too that all top positions were being given to Germans. Coincidentally King Frederik V's first wife, Queen Caroline, daughter of George II founder of Göttingen University, was born in Ansbach.

An organization separate from the university was established. The Royal Botanical Institution was created and G.C. Oeder was appointed Royal Professor of Botany. At twenty-four

Trondheim, Norway (I) 1758–1759

Trondheim Cathedral as it stood c1800, only a small section of the building was usable.

years of age he was given the responsibility, under Count Moltke, of planning the Royal Institute, necessitating an extensive tour of similar establishments and botanical gardens in Holland, France, Germany and England.

A notable exception in his tour was Sweden, where Linnaeus was a popular professor and had re-established Rudbeck's botanic garden in Uppsala. It would have been a rather long journey for an enthusiastic young man eager to establish the King's Garden and he had been to Holland, where Linnaeus had spent some years. Or perhaps he had been indoctrinated by Professor Haller, the most conspicuous botanist/biologist in north Germany, Switzerland and Denmark, who was totally against Linnaeus' ideas considering his 'System' artificial.

The plans for extending and updating the existing records of the country's natural resources came within the new institution. Oeder had to purchase books for the new library and source illustrators and copperplate makers for the printing of the records.

Our main characters have now taken their places on the mid-eighteenth century Trondheim 'stage', with its' backdrop of an almost derelict cathedral of ancient historical and religious significance and in the wings, impressive timber buildings, many new at that time (some still in use in the twenty-first century), demonstrating the affluence and scholarship of the supporting cast.

The first edition of Gunnerus' First Pastoral Letter published in 1758 shows Schøning's signature.

Trondheim, Norway (I) 1758–1759

This portion of the Trondheim Seal indicates the importance of the Nidaros Cathedral in the affairs of the City. The Bishop raises his right hand in Benediction (Blessing) whilst in his left hand he holds a shepherd's crook, a symbol of his office. 'Hyrde' means a herder, or gatherer of animals as in shepherd - a good shepherd who leads his flock of sheep and guards them. So the bishop cares for his flock in the green pastures and on the barren, rocky mountain side. The crook was useful to reach sheep in difficult situations. The shepherd would hook it round the sheep or lamb and gently carry it over his shoulders back to the fold.

Bishop's Letter

Bishop Gunnerus soon applied his thorough, broad-based education and previous experiences in writing, debating and lecturing to his new position. Within weeks of his installation he wrote an encouraging pastoral letter to the diocesan clergy clearly demonstrating his own firm grounding in the orthodox Christian faith. The clergy should scholastically cultivate the world around them, the place where God had placed them in His world and socially provide a Christian example.

Gunnerus' First Pastoral Letter was printed and published in Trondheim in 1758. A facsimile edition of "Hans Opvækkelige HYRDE-BREV" (His Stirring Pastoral Letter), was published by the University Library of Trondheim in 1997. The original letter chosen for reproduction was a copy bearing Schøning's signature, possibly a personal gift from the bishop.

Published in the Dano-Norwegian language of the time and in Gothic script, I personally found translation difficult. By studying a modern English Bible translation of the easily recognisable biblical references given I was moved by the sentiments conveyed. The few Old Testament references were more difficult to ascertain for chapters and verses were not consistent with more recent Hebraic translations.

Gunnerus had taken a Pauline Epistle pattern and had so aptly applied many quotations from Paul's Letters to Timothy – and what could be more appropriate – the apostle to his disciple, the father in Christ to his adopted son? In every extract presented, in epistle or gospel, I could read some intimation of the natural world around the lives and activities of the bishop's flock – the landowner, the vineyard, the good things, the store-room, treasures, light, shipwreck, the good deposit....

> What you heard from me keep as the pattern of sound teaching, with faith and love in Christ Jesus. Guard the good deposit that was entrusted to you – guard it with the help of the Holy Spirit who lives in us. (2 Timothy chap.1, verses 13/14.)

Ecology is not new! This 'good deposit' is still entrusted to us.

The germinating seed of a natural scientific society was well and truly sown and later enhanced by an extended edition of the letter in German, the language of the Danish court and universities where Gunnerus had studied and lectured.

In exhorting his clergy to search and explore the God given treasures around them he gave examples of those who had done this. He presented a long list of the names of many of their colleagues who had written and published articles on some aspect of natural science.

Before pronouncing the familiar words of St Paul, which we use as Benediction, he concluded his writing with the stimulating words, quoted from Paul's epistle to the Romans

> Never be lacking in zeal, but keep your spiritual fervour, serving the Lord. (Romans 12, verse 11.)

Here we have one clue to *Gunnera* for Gunnerus zealously practised what he preached. (Several species of the genus are zealous, fervent growers.)

Sheep are a familiar sight in temperate climates, as they roam over hills and pastures.
It is too hot for these animals in the tropics because their wool is thick though it provides one of the best insulation materials in the world. They are good meat too - mutton or lamb (if young). But we call them 'silly sheep' for they are easily led, where one goes the others follow, just like people (the flock).

The shepherd is ever watchful. In the Bible Lands he went ahead of his flock to make sure the way was safe and the sheep followed. The Good Shepherd – as in Psalm 23.

Trondheim, Norway (I) 1758–1759

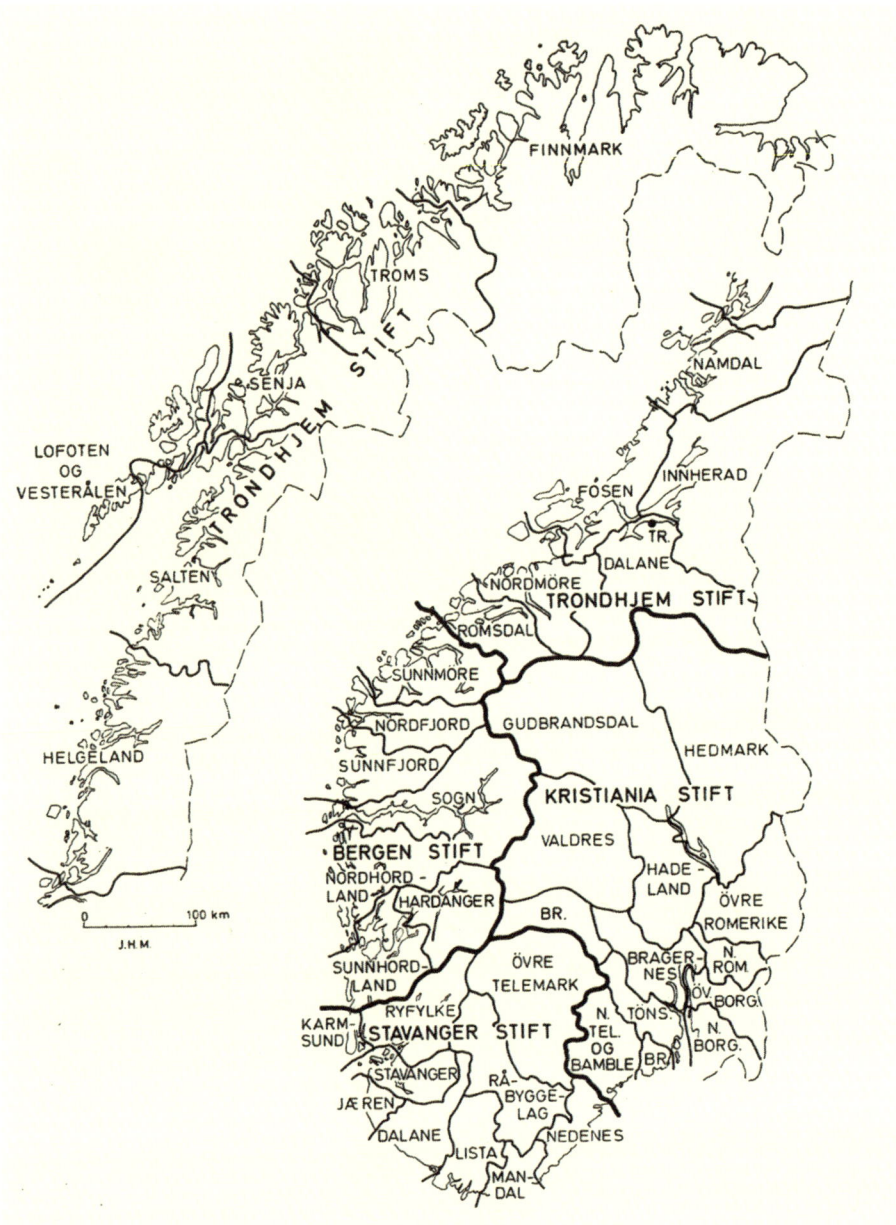

Map of c. 1665 showing:

Diocese ——— a Bishop 'stift'
Deanery ——— an area of several parishes – a Dean

Sogneprest – a vicar, parson in charge of a parish under a Dean.

4

North Norway
1759–1760

First Pastoral Journey

As the snow receded from the distant mountaintops and daylight lengthened, excitement mounted in Trondheim in May 1759. There must have been increased activity during the last few weeks in the boatyard at the mouth of Nidelva. A new ship, ordered by the bishop, was anchored in the harbour within sight of the cathedral and final preparations for departure were complete.

It was the largest cruise ship seen on the fjord, with a crew of twenty men, including eight pairs of oars. The stateroom provided ample accommodation in which Gunnerus could read and write in an undisturbed environment. Local clergy and friends would accompany him and the long, slow journey would provide the quiet time to discuss affairs of church and the developing scientific society. Sometimes his housekeeper-sister Stine, and one or other of her daughters would

North Norway 1759–1760

Gunnerus' first northern journey.

Fish drying over a rack in the fresh salty breeze.　　　　　　　　　　　　AHC

join the company ensuring comfortable, pleasant travelling - the sea permitting!

The party travelled northwards from the mouth of the Trondheimsfjord. Trade was booming along this northwest coast in lumbering and dried fish for export. The catch was brought ashore, cleaned, gutted and then suspended over horizontal wires supported by crossed poles in the ground, to dry naturally on the rocky shore. The completed structure appeared like the roof of a building, the staves forming the gable ends and the grey silver scales of the fish the tiles.

In Gunnerus' own writing, one can still read the record of that journey in 'TRONHJEMS BISKOP VISITATSPROTOKOL 1732 18/8 TO 1770 7/9" (page 370) in the Statsarkivet, Trondheim. *(The Bishop of Trondheim's Visitation Register)* The writing is frequently untidy, difficult to read and blotchy in places – evidence of an unstable surface, as on a pitching ship. Easily discernible are the times when a new quill is substituted – a clear reminder of the vast changes in calligraphy and stationery materials over the years.

The first entry refers to Sunday, 16th May 1759 when Rasmus Jacobsen of Saltdalen was confirmed. The bishop's vessel would have anchored in Saltfjorden within the Arctic Circle and folk would have gathered in a church in the Bodø area to welcome the bishop on his first visit to the north. The parish of Salten, like all except two parishes (Romsdal and Nordmøre) in this extensive diocese, was bordered by Sweden and included the narrowest part of Norway, only three miles wide from the head of Tysfjorden, at Hellemobotn, to the Swedish frontier.

Carl Linnaeus, a young medical student from Uppsala University in Sweden with a penchant for natural science, visited this area on his Lapland journey in 1732. Walking, climbing and clambering over uncharted boggy wastes, making more progress in boats on rivers and extensive Swedish mountain lakes, (except the boat had to be carried to avoid rapids and waterfalls) and sleeping in a Lapp portable tent (a circular tent of reindeer skin stretched on birch poles). A Lapp guide, in his insulated, dried-grass filled, reindeer skin boots, would spring along an unmarked way sometimes leading the young foreigner astray.

Leaving Saltfjorden, Bird Rock near Bodø.
AHC

Linnaeus in his journal, 'Lapland Journey' records that the final guide ran and left him when he saw the Norwegian Sea in the distance far below, for this meant easy access to spirits and tobacco. Linnaeus stayed some days in Norway at Sørfold on the fjord parallel to Saltfjord.

Linnaeus' way over from Swedish Lapland to Norway. 1732.   AHC

Gunnerus was beginning to hear more of Linnaeus in the course of his discussions with Oeder, Suhm and Schøning. In 1737 Linnaeus had been elected the first president of the Royal Swedish Academy of Science. The Trondheim friends hoped to establish a similar association in Norway, which would eventually lead to the establishment of a university in Trondheim. Linnaeus had been appointed Professor of Botany at Uppsala in 1741, a seat he held for more than thirty years.

A typical fjord village in the early setting winter sun (about 1400 hours).

By 21st May 1759 the bishop's party had reached the island of Hamarøy in the great Vestfjord, which parts the large group of Lofoten Islands from Norway and its fringe of islands (the Skerries). There had been visitations on each of

The jagged peaks of the Lofoten 'wall'.
AHC

In Pontoppidan's Bishopric near Bergen.

the three previous days and more confirmations recorded. Gunnerus had found at Lødingen, on the island of Hinnøya, some confusion in the examination of confirmation candidates, as the 'old method' of Pontoppidan's catechism was still in use, indicative of the slowness of communication in the remote north. (Erik Pontoppidan, a Dane, born on the island of Fyn in 1616, had been the Bishop of Trondheim 1673–78. He had Luther's catechism translated into the Sami/Lapp language. Hence the reference to the 'old method'.)

Gunnerus would come in contact in Copenhagen, both as a student and lecturer, with another Erik Pontoppidan, great nephew of the deceased Bishop of Trondheim. A great exponent of Pietism he was commissioned by King Christian VI to compile a new explanatory catechism published in 1737 entitled, 'Truth for the God-fearing', abbreviated in 1771 by Saxtorph and used for nearly 150 years. On the ascension to the throne of Frederik V in 1746, Pontoppidan was 'banished' to the bishopric of Bergen to dispel the pietistic influence in the Danish Court. He remained in Norway eight years as Bishop of Bergen before being recalled to a university position in Copenhagen. He is remembered for the enormous extent of published works, ranging from Danish church history to an attempt at Norwegian Natural History (1752–53) and the 'Danish Atlas'.

The bishop spent the next four days in the Ofotfjord and Tysfjorden. The head of these fjords almost reached the Swedish frontier and Gunnerus found fifty Finnish families resident in this locality. They had come over the eastern wet wilderness of Stora Sjöfallet to the North Sea for the more bountiful supply of fish, eventually building homes there and farming the narrow coastal strip beside the fjord. They had their own missionary who had been trained at the Lapp Teachers' Training College in Trondheim.

Moving northwards from Tysnes and Korsnes, again crossing Ofotfjorden, Gunnerus would leave behind a small area that in later years would become Norway's first National Reserve, to reach Evenskjær on 25th May. They passed through Astafjord, visiting Astafjord Church 28th May and sailed into the mountains, or so it would seem for this fjord, Gratangen, had only a very confined coastal shelf. At the head of the fjord there are still three churches marked on the map but there appear to be very few farms or habitations even today. It must have been a gathering place for a market and special festivals. In other places it is found that a new church had been built when the first became too small as a gathering place for the growing population, or the roof needed repair, or some wealthy Lapp wanted his own church…Two or three tracks lead up forest covered valleys – perhaps this was a timber exporting port centuries ago.

> (I remember arriving in the sunlit early morning hours at the hotel at Gratangen, having found all hotels in Narvik full. I gasped when I saw the scene from my balcony for it was an exact replica of my favourite Geirangerfjord)

Tromsø, one of the most northerly cities in Scandinavia, was reached on 30th May. In May 1759 and still in 1899, when Alfred Heaton Cooper sketched it, Tromsø was an undisturbed fishing town, boat building and trading crossroads. Though

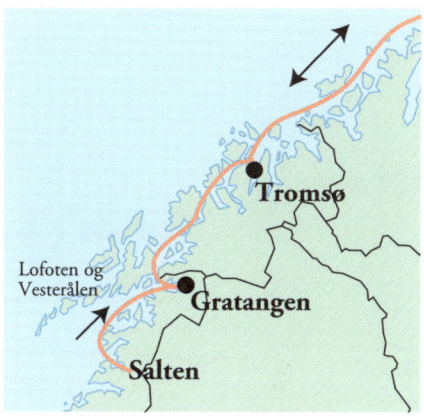

First Northern Journey continued.

Geirangerfjord

Tromsø harbour   AHC

protected by several islands from the Atlantic Ocean the area enjoyed the benefits of the warmth carried by the North Atlantic Drift, the current carrying warm waters from the Gulf of Mexico to moderate the climate of coastal Arctic Norway.

The most impressive Tromsø landmark today is the modern cathedral. The town was devastated during World War II but out of the ashes a prosperous city has developed with "the white cathedral" dominating the scene in a simple unpretentious manner. It stands with its steeply pitched roof like a great triangular iceberg – the apparent three sides suggesting the indomitable power of the Trinity. Four fifths of an iceberg cannot be seen – the unseen foundation of this 'Cathedral of the Arctic' would include the work and influence in untold ways of hundreds of people through centuries, men like Johan Ernst Gunnerus.

The script in the next entry in 'Visitatsprotokol' is most irregular, obviously written during a rough passage across the Altafjord. Gunnerus records preaching on 8th June at a chapel eight miles from the farm where the party was entertained. No further record appeared until 19th June when Kjøllefjord Church was reached. The northern mouths of Porsangerfjorden and Laksefjorden had been traversed.

Disembarking a few miles from Kjøllefjord the bishop transferred to an open boat to negotiate the narrow fjords and reach Tana Church built where the Tana River joins the Tanafjord. He visited Lapp encampments, only accessible by a small rowing boat, for he was anxious to learn more of this traditional culture and the nature of the area with which

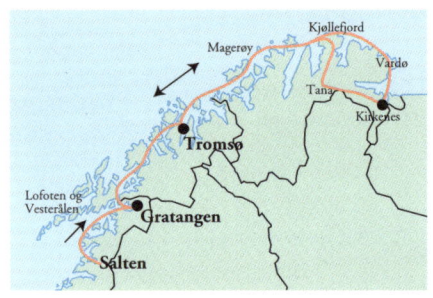

First northern journey continued.

these people were so in tune. He would also need to make contact with the missionary teachers who had studied in Trondheim. In the account of 21st June there are certainly many Lapp names recorded as pastoral lay workers.

Gunnerus then travelled south overland beside the wide, raging river supplied by the perpetually snow covered mountains of northern Finland, to Tana Bru (bridge). There the group turned eastwards over the packhorse track to Varangerfjord reaching Nesseby Church on 23rd June and rejoined their vessel.

On the overland journey Gunnerus had been able to observe the wild life and to collect some plants for a garden and some specimen for his developing herbarium in Trondheim. Some names were new to the country and were identified only after being forwarded to Professor Oeder in Copenhagen – *Veratrum album v. lobelianum; Primula sibirica.*

These two species are still rarely found anywhere else except in North Finnmark, the coastal area around Vadsø and Vardø, the old outpost fortresses near the wild Russian border.

The Norwegian Nyserot, (lit. trans. 'sneeze root') *Veratrum album,* sometimes falsely called hellebore, is poisonous. It is a fairly large plant growing to 60–120 cm, flowering June to August, with the long sunlit days of the arctic summer.
The flowers grow in spikes with florets of six petals of golden green. The deeply veined oval to elliptical leaves, growing directly pleated from the stem and overlapping, are particularly interesting but they have nettle-like hairs beneath.

The sun at midnight, June, in Swedish Lapland (Abisko).   AHC

Veratrum album. N. Nyserot

North Norway 1759–1760

Primula nutans
N. Finnmarksnøkleblom

Primula sibirica, now included with *P. nutans,* is the Norwegian Finnmarksnøkleblom, the name stressing its habitat. 'Nøkleblom' is commonly called cowslip but 'nøkle' means 'key' so way back in time I guess the flower was recognised as the key opening the summer pastures for the seter girls and the goats and cows, released from their dreary winter imprisonment.

Flowering June to July it is 4 to 25 cm in height and has 2 to 4 pink/mauve petalled flower heads on delicate stems. The plant likes a sandy soil so it is frequently found in the short turf near the northern sea.

What a companionable treat it would have been if Oeder could have been in the group. It would have been a convenient, comfortable manner in which to survey this remote area and Gunnerus would have had expert tuition in this aspect of natural science, which was gaining his particular interest. But Oeder had been ordered to Copenhagen.

After an absence of five years from the university, no material or report had been received from Oeder. On 27th April 1759 a despatch was sent to him in Trondheim, 'Start immediately, no more delay!' He did, only to find that details he had forwarded had been covetously retained by the professor of medicine instead of being forwarded to the copperplate engraver. The work on the edition of 'Flora Danica' then progressed in Copenhagen and Oeder was able to resume his research in Trondheim.

Gunnerus reached the eastern outposts of Vardø and Vadsø, old fortresses near the Russian border. I speculate he may have travelled a few miles further eastward along the coast of Finnmark to what may have been an isolated church on a promontory of land accessible by sea and land.

This is the meaning of the name Kirkenes, the most important Norwegian town along the north coast today. The immediate area developed rapidly around 1904–06 when iron ore was discovered in the area, similar to the deposits found in Swedish Kiruna which had to be exported on the specially built railway through Narvik, an island sheltered North Sea, Norwegian port. Kirkenes is somewhat protected from the vicious Barents Sea by the Varangerfjord. Although the Barents Sea is a good fishing ground at certain times of the year, the locals refer to it as 'the devil's dance floor'.

Lapp kåta (reindeer skin over birch poles, portable tent)

Fish drying on the rack and one reindeer nearby. It is difficult to see the man in his traditional blue costume, a purposeful camouflage.   *AHC*

[I endorsed the epithet from personal experience after crossing to Svalbard in June 1995.

In contrast to the anticipated St Hans, midsummer celebrations, when the sun never dips below the horizon giving twenty-four hours daylight, we encountered a bronco-riding storm of gale force 13. At the height of the storm the emergency alarm summoned passengers and crew to lifeboat stations. Fear of the deep, dark ice-cold water with yawning, frothy peaks was only felt afterwards when tension was visibly relaxed on being told the alarm was due to an electrical fault on the bridge.

There could not have been an electrical fault on Gunnerus' bridge in the eighteenth century but he certainly experienced such storms]

The Bishop turned homewards to Trondheim from this north-eastern boundary of his bishopric. On 5th July 1759

North Norway 1759–1760

the ship paused at Kjelvik at the mouth of the Kjøllefjord, which had previously been visited on 19th June. The visit to the island of Magerøy on 7th July was recorded with a visit to the settlement now known as Honningsvåg. The sea must have been more pleasant as they turned southwards round the northerly corner of Europe (now the North Cape-*Nordkap*-Tourist attraction) reaching Hammerfest five days later.

Church visitations were made on 13th July at the small island of Loppa in the outer skerries and on 15th at Skjervøy. Sunday, 17th July was especially memorable for the Bishop made a detour into Lyngen to visit the area for which a translation of the Bible in Lapp had been produced by the college in Trondheim.

The homeward journey was resumed, passing through Tromsøfjorden and then visiting churches and communities not reached on the outward journey: – Sortland, by a narrow sound Steigen, Nesna and the lovely island of Dønna, anchoring at Dønnes.

Gunnerus could not help but have admired the beautiful skyline on the island of Alsten, between the fjord and the mainland, of the mountain range of the 'Seven Sisters' as he sailed across to Sandnessjøen. This very name implies a peninsula of sand in the sea, providing easy anchorage for boats and sailing ships. When I first visited Sandnessjøen in 1960 on a coastal steamer it was accessible only by sea but by the time of my second visit in 1970 it was possible to reach the growing town by road and ferry but 'the Seven Sisters' remained unchangeably aloof.

Drying hay over wires streched between moveable poles. Every square inch of meadow had to be conserved for winter fodder.   AHC

One of the most radiant days I have ever spent in Norway was on the island of Dønna on a luminous day in July 1970. The placid brown cows were leisurely ruminating in the buttercup strewn bright green pastures beneath a clear azure sky. Behind me was the white painted church, protected by a few low silver birch trees with shimmering pendant leaves and before me the distant snow covered mountains.

On 22nd August 1759 the Bishop recorded his visit to Alstahaug Church on the southern tip of Alsten Island. Two days later he was in Brønnøysund and in this manner the journey was completed, making at least five more visitations and finally reaching Trondheimsfjorden on 7th September 1759.

It is interesting to note that the Hurtigruten, the express coastal vessels, daily plying the Norwegian coast between Bergen and Kirkenes since the mid nineteenth century, mainly follow the route Gunnerus took through the skerries and fjords, a century before. The safe passages through sounds and open fjords must have become known from centuries of navigational skill developed by the Vikings – the men of the creeks.

Express Coastal Vessel in Trondheimsfjord. March 2001.

Two pages from AHC's sketchbook provide a wide-angle view of Sandnessjøen with the Seven Sisters Mountains on the left. The church stands prominently on the nes giving easy boating access.

5

Trondheim (II)
1760–1763

The Learned Society of Trondheim

The dark days of winter, brightened only by the candle lights of Advent and the church and family festivities of a snowbound Christmas – *Jul,* pronounced 'yule' as in Yule log – were enlivened by long, fervent discussions in a cosy library.

Gunnerus had been on his journey to the unfrequented north of the country, Suhm had no doubt all the latest news from Copenhagen and Schøning was deep into his research on the history of Nidaros Cathedral. Friends would join this triumvirate for knowledge and entertainment, airing their questions and ideas, stirring memories.

They were frequently joined by Oeder, who had returned to Trondheim from Copenhagen in the winter of 1759, to finalise his botanical assignment and no doubt anxious to see Gunnerus' collection of specimen.

Advent Candles.

An additional candle is lit on each of the four Sundays of Advent and then a special one on Christ's Feast Day itself.

This modern safe example was spotted at Uppheim Church, near Voss - an electrically supplied system, a day after the second Sunday in Advent 2001.

Trondheim (II) 1760–1763

Berg-i-Strinden, (now Berg). Gunnerus had a garden here and lived for a time in this parsonage.

Another welcome, regular visitor was the Trondheim physician, *Dr Robertus Stephanus Henrici,* a particular friend of Gunnerus and Oeder, as he shared their interest in natural science. Like many doctors he had his own herb garden at Stene-i-Opstrinden (now Steinan) and coached Gunnerus in dissection and basic taxonomy. Gunnerus had heard and read of the work of the brilliant Linnaeus, professor of botany at Uppsala University in Sweden, but in Oeder and Henrici he had 'on the spot' tutors.

Bishop Gunnerus had become acquainted with most of his clergy and was able to disseminate through them a basic philosophical study of Christianity and natural science. He desired to form a society, as he had experienced in Denmark, Jena and Halle, where deeper study and discussion of developments could take place amongst the select educated leaders. Such a society would enhance earlier teaching and publish literature and poetry, articles on physics, natural science and history and provide a platform for public speaking and debate.

The time was ripe to launch such a project for Gunnerus himself had experiences to record of his arctic tour and Schøning and Suhm produced papers on historical and literary matters they researched. In 1760 the first scientific and literary society in Norway was formed, *Det Trondhiemske selskab (The Learned Society of Trondheim).*

The list of founding members includes Professor Oeder and Dr Henrici together with Gunnerus, Schøning and Suhm, several clergymen, prosperous merchants and civic leaders. Membership was by invitation and the fee, the donation of two books for the establishment of the society's library, was instigated later.

The following year saw the publication of the first proceedings of the Trondheim Society of Sciences and Letters. Volume I included articles by Gunnerus on seabirds and minerals he had observed in Nordland and Finnmark; some general notes on world history by Suhm; an account of bad weather and crop failure in the area of Trondheim and a business report of public activities within Norway by Schøning.

1761 saw the publication of the first volume of 'Flora Danica' in Copenhagen under the editorship of Georg Christian Oeder. Volumes II and III followed in 1762. Martin Rössler's meticulous botanical drawings were transferred to copper plates by his father, printed and then individually coloured by hand. This was such a painstaking, time consuming process that a cheaper, uncoloured edition of the book was produced. The work was subsidised by the Privy Purse because it was a state commission.

The 'black and white' edition was distributed, as soon as possible, to all the bishops for them to publicize in the cathedral schools and encourage their clergy to observe and collect further information. An original copy is still to be seen in the archives of the Royal Norwegian Society of Sciences and Letters inscribed,

SCHOLA·CATHEDRALIS·NIDROSIENSIS.

The detailed illustrations are unique and remain accurate, diagrammatic descriptions of flora, a standard not surpassed

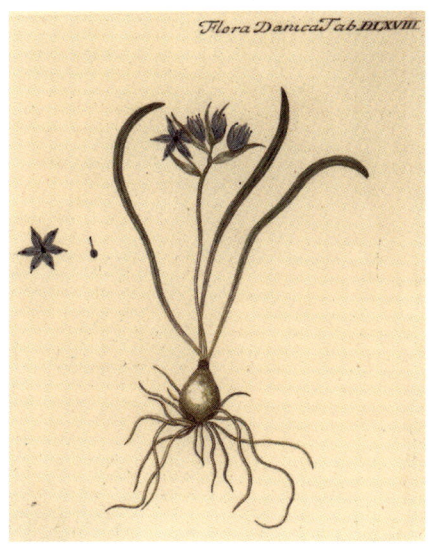

Scilla bifolia Oed.
From Flora Danica. Exactly reproduced on the authors 'Flora Danica' wineglass bath. (Page 135)

by photography. Oeder edited the first ten parts, containing plates 1 to 600, which were available between 1761 and 1771. Several full copies were circulated in Trondheim.

> (This was not the first 'Flora Danica' for one had been produced the previous century by Simon Paulli (1603–1680), a German botanist and physician who was professor of medicine at Copenhagen)

Gunnerus must have been inspired by and gained knowledge from his friend's work. From this time onwards he gradually paid more attention to botany, providing systematic details and preparing his own herbarium. Previously his special interest had been zoology as evidenced on the First Day of Issue of stamps in October 1970 depicting Norwegian Zoologists. The set of stamps includes an engraving of Gunnerus.

FDC Norwegian zoologists issued October 1970. Gunnerus on upper left stamp.

Gunnerus' *kopibøker (-er ending denotes plural)* include correspondence with several Danish friends regarding available literature about natural sciences. He enquired about the possibility of Professor Oeder obtaining a microscope for him, as he would like to use it in his studies with Dr Henrici. The idea was developing that he should write a Norwegian flora for he was fascinated by the different dialect names for the same plant, the name often indicating its local use.

Råshult, birthplace of Carl Linnaeus.

Carl von Linné

On 24th April 1761 Johan Ernst Gunnerus wrote his first letter to the leading natural scientist of the day, *Carl von Linné*, the professor of medicine and botany at Uppsala University, the oldest university in Sweden (1477)

Baptised Carl Linnaeus in the church at Stenbrohult, where his maternal grandfather was Rector and his father curate. He was born nearby at Råshult in the gentle, pleasant county of Småland in southern Sweden, in 1707. His birthplace was a simple residence built by his father with a specially created garden, fenced within the surrounding supporting farmland but most of his childhood was spent in the Rectory at Stenbrohult where his father succeeded his grandfather. The church remains on a promontory in Lake Möckeln and one can still experience the joy and freshness of nature's bounty as the youngster did with his amateur botanist father three hundred years ago.

Carl attended the cathedral school at Växjö, some thirty miles away. Fortunately a wise headmaster (Rothman) persuaded

Växjö Cathedral.
The building adjacent to the left of the church was the Grammar School Linnaeus attended.

Trondheim (II) 1760–1763

father Nils Linnaeus that the frequent truant, who longed for the company of birds and flowers, was more likely to succeed as a physician (including botany) than as expected, a clergyman.

Linnaeus spent a short period at Lund University in his twenty-first year, before transferring to Uppsala University where he remained for the rest of his life, except for the necessary period abroad to complete his medical doctorate. The years in Holland provided the opportunity to publish some of his earliest botanical writings, for there printing and publishing were more established than in Sweden. There too he was granted the generous support of the Anglo-Dutch merchant, a keen horticulturalist, George Clifford. He was able to visit France and England before returning to Sweden in 1739, never again personally travelling abroad but exploring the far corners of the world through his 'disciples'.

In 1757 he was ennobled, Carl von Linné, and as such he was addressed by Gunnerus when their correspondence commenced in 1761.

The bishop's summer visitation was in the southern part of his diocese, across the Trondheimsfjord to Stadsbygd on the most southern promontory of Trøndelag, the deanery of Fosen. He remained mainly within sheltered waters and easy distance of his headquarters, though mention is made in the records of time spent in Ørlandet. This was the flat, fertile area around ancient Austrått, between the mouth of the Trondheimsfjord and the open sea.

Twilight is early (1400 hours) in winter around the fjord and the gradual drop in temperature causes a rippling breeze.

Trondheim (II) 1760–1763

Bishop Gunnerus' pastoral ministry continued in its various ways, his dynamic enthusiasm being evident in all his interests. In early June 1762 he commenced his second northern journey reaching Alstahaug on 12th June. At the southern foot of the Seven Sisters Mountain peaks, south of Sandnessjøen and Dønna, on the main seaway to the north, Alstahaug was a famous focal point. The talented poet, half-Scots clergyman, Peter Dass was vicar here and recorded his experiences in verse, 'Nordland's Trumpet'. He was so revered that it is said on his death in 1707 the local fishermen sewed black patches on their sails. His son, Anders Dass had studied at Oxford before taking over his father's parish of Alstahaug. Anders had married Rebecca, the only daughter of Lorentz Mortensen Angell and so was uncle by marriage to Thomas and Lorentz Angell, the sons of Albert, his wife's eldest brother.

At the time of Gunnerus' visit it would be a most hospitable farm and vicarage and today the church has been restored. The name of Dass was unknown to the occupant of the old vicarage when I searched in 1968 and she could not understand why an English lady should want to find Alstahaug at the end of the twentieth century.

The Visitation included Bodø, Hammerfest (27 June) and Tromsø (4 July), the dates given would suggest the bishop visited Tromsø as he turned south, Hammerfest being the more northerly point. Out to the west, in the open sea, time was spent in the Træna group of islands with a single church, on 9th July 1762. The earlier part of this record was in a different script suggesting the work of an assistant. The resumption

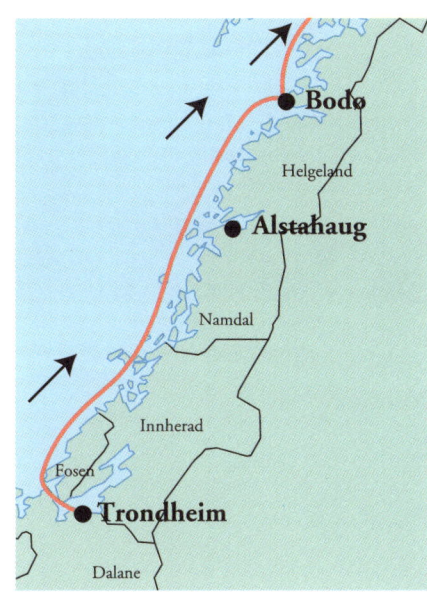

Second Northern Pastoral Visit 1762

Romsdalfjord in summertime.

Trondheim (II) 1760–1763

The well-drained land round Trondheim has been well and prosperously cultivated for centuries.

Gunnerus' Microscope

of the bishop's personal writing is indicative of his relaxing voyage home.

Awaiting his return was a letter from Linné, dated 4th July 1762 in Uppsala. It was a short, friendly reply apologising for the delay in response but encouraging the doctor of divinity to use his God given talents carefully and well.

Before sending a further long, enquiring letter to Linné, 9–11 October 1762, Gunnerus had written to Oeder in Copenhagen in September detailing his recent northern visit. He had collected some beautiful and unusual plants, some to be planted in his botanical garden and some for his herbarium.

Preparations were in hand for the second volume of the proceedings of the Trondheim Society and Schøning published a well-researched history of Nidaros Cathedral. (The volume was available mid-1763)

In January/February 1763 the longed for microscope was safely delivered to Gunnerus in Trondheim. Oeder had obtained the 'aquatic microscope with four magnifiers' from John Cuff optician of Fleet Street, London, the previous March. Finn-Egil Eckblad in his article, *Biskop Johan Ernst Gunnerus (1718–1773) og hans mikroskop*, in Blyttia 42.1–5, 1984; wrote it was 'aquatic' because the object under inspection could be placed in water on the glass plate between the mirror and the lens, all three circular fields being adjustable on a vertical rod. It cost three guineas – 15 Norwegian riksdaler. Some mystery surrounds the microscope's appearance, disappearance and subsequent reappearance. Now it rests safely in the Gunnerus Collection in the museum of Natural History and Archaeology in Trondheim. There is no doubt that Gunnerus used it and derived benefit from it for he mentions it in correspondence with several friends.

(Gunnerus' solar microscope, also listed in his effects, remains an unsolved puzzle.)

The annual summer visitation of the bishop in 1763 was different from previous visits for it was inland. It included the inner portion of Trondheimsfjorden, adjacent Beitstadfjorden and the extensive Lake Snåsa so the boat could have been used as base part of the time. The eastern boundary was the mountainous Norwegian/Swedish divide with farm tracks up narrow valleys, too interesting to hurry over. Gunnerus was now deeply interested in botany and in this mountain terrain he found different flora. He knew the people lived closer to the land, land that had been farmed by generations of the same family, unless disturbed by the Swedes. The benefits and uses of the herbs and flowers had similarly been passed on by word of mouth and experience. There was much knowledge to be gleaned from the local parishes and to be recorded.

During this period the bishop wrote many articles to his clergy about natural science hoping they would collect further items of interest and publicize the wonderful revelation of God around them.

The two churches at Sakshaug at the head of the Trondheimsfjord show the diligence of long gone ancestors who had laboriously cleared the ground of stones and rocks brought down in the Ice Ages.

Trondheim (II) 1760–1763

Innherad, Dalane, Nordmøre and Romsdal

6

Trondheim (III)
1764–1766

A golden period followed – that rare time of life when interests, projects and responsibilities successfully meld together and progress. Gunnerus was encouraged by a letter from Carl von Linné dated 12 March 1764. It is not known exactly when the bishop received this correspondence but he replied with further details on 19th May 1764. It is important to be reminded, in this age of instant communication, that in the eighteenth century letters could be delivered only irregularly by the state postal service or when some friend or acquaintance was going in the desired direction and transportation could take weeks. There are many interesting biographies veiled in 'postmen' of the time, students of Linné, clerical friends of Gunnerus, mutual aquatainces of the wealthy Trondheim merchants and reliable travellers who had assisted in this procedure.

The style of Linné's letter to Gunnerus is strange, almost amusing, until we recall the date and recollect that letters were

Trondheim (III) 1764–1766

then very rare. Not only were there a comparative few of the total population who could read and write but also writing itself was an expensive, laborious task. Paper was being handmade in a few small local family concerns and ink produced from lampblack (a form of carbon), submerged in a colourless quick-drying viscous fluid like boiled linseed oil or the clear, sticky liquid from oak-galls. The small sharp knife used to cut the end of a quill (feather) at an acute angle, to form the writing instrument, is still called a penknife.

> 'Carl von Linné, knight, sends his excellency, Johan Ernst Gunnerus, the greatest bishop I know, the Pliny of the north, many greetings.'

So commences the 1764 letter. He commends Suhm's writings in the second volume of the proceedings of the Trondheim Society. The praise for Gunnerus' writing commends his elegant, concise and detailed descriptions and Linné declares he could not have done better himself.

> 'I felt I could see your specimen with my own eyes and feel them with my hands', he wrote.

After various specific references he concludes with further compliments –

> 'You alone reveal both of God's books – His written word and His world of nature. At the same time your enlightenment is like a great northern light more than any other bishop in Europe. May the great triune God long keep you sound and healthy!'

Gunnerus replied in a similar manner, grateful for the encouragement given by the master and expressing the wish they

A fjord winter sun set (1400 hours) over Hardanger.

could meet, *'Oh! If only Uppsala had not been so far from Trondheim.'* In the course of the long letter, including further details from other sources and the subsequent renaming of some specimen, the bishop indicates the preparation of the first part of his 'Flora Norvegica' for the Trondheim Society. He did not intend to travel far on his summer visitation that year but spend the time in the surrounding deanery of Dalernes, within comfortable travelling distance of his home and plants.

Dalernes was a unique area for it included everything the other *prosti* did in the extensive diocese, except an open coastal area. In addition it incorporated the ancient city of Trondheim and the most important copper mining area in Norway, Røros.

The northern boundary of Dalernes stretches across the Trondheimsfjord including the mouths of three important rivers that have been lifelines to the area through centuries: – (east to west), the River Orkla in Orkdalen; the River Gaula rushing through Gauldalen from the watershed where it has been parted from the mighty Glomma River in Østerdal, the Great Eastern Valley; and the River Nid, fed by the large Selbu Lake into which several streams flow from the Norwegian/Swedish mountains.

Each valley has been a significant pathway for more than a millennium. They were worn pilgrim ways, with rest houses (cold harbours) like Kongsvoll, through the neighbouring diocese of Christiania by way of Hamar and Ringebu, up the kindly Gudbrandsdal valley and over the harsh Dovre plateau to the shrine of St Olav. Such routes now carry railroads and highways, following the age trodden courses between Oslo and

Uppsala, Sweden.

The old Botanic Garden and house restored by Professor Linnaeus. The spires of Uppsala Cathedral are seen in the background.

Dalernes

Trondheim (III) 1764–1766

Orkdal Fjord

RØROS c1700 showing the church built in 1650 and the copper miners' homes in the lower part of Kjerkegata and the school and larger houses nearer the church.

the north. Orkdalen still has a westerly branch leading to Surnadalsfjord, in the *prosti* of Nordmøre, and so to the open sea.

Røros remains a capsule of eighteenth century history. In the white stone church nestling on the hillside, set above the great residential houses of the past, one may experience the heart-warming spirit of the culmination of Gunnerus' teachings and example in the theological philosophy of the Reformation and natural science. The church's large, clear-glass leaded windows admit God given light showing the white painted pulpit set very high above the altar, indicating the high esteem given to God's spoken word. In the winter the same windows would reflect the lights of many candelabra against the darkness of an outer world often glittering with frost and snow. (Røros is frequently recorded as being the coldest place in Norway)

Bishop Gunnerus did not actually preach in this church, for it was only completed in 1784, but he would know its close neighbour, the smaller wooden stave church of 1650. The wooden church remained standing for a few years after the completion of the 'new'. Gerhard Schøning, a co-founder of the Trondheim Society, did produce one of his characteristic drawings to illustrate the position of the church in relation to the original mining village.

Copper mining commenced here on 'Bergstaden' *(mining town)* and the wooden smelting hut was in operation from 1644. Legend recounts that a reindeer, in its death-throes, kicked up a shining nugget of pure copper, on the gravelly hillside. This was recognised and so prospecting successfully commenced. A town with two parallel main streets of wooden, turf-roofed houses, quickly developed. The king, Christian IV, the prolific builder, owned all mining rights but these were handed over to an adventurer, Joachim Jürgens, in settlement of the king's debt. 'Jürgens', Norwegianised to 'Irgens', soon gained control of the growing business.

The surrounding pine forests would supply the wood and charcoal for smelting. The small farms in the area were mostly self-supporting, with cattle, sheep and goats to provide dairy products, meat and wool. The men folk would be occupied in timbering and the hazardous, laborious task of charcoal manufacture. The whole coniferous, soft-wood tree was felled and stripped of its branches in summer or autumn and stacked until the early winter frosts hardened the ground and the snow made it easier to drag the timber by horse drawn sleigh to winter stores near home. Grains had to be imported from the fertile agricultural district around Trondheim and many small flourmills would be found over a rushing beck, complete with the famous, special grinding stones from nearby Selbu.

With the development of the Røros Copperworks there was a surge in demand for timber not only for smelting but also for building. Accommodation had to be provide for workers seeking employment, officials and public buildings. Brush wood was needed too in the pre-dynamite days for breaking stones to retrieve the copper from the ore. A fierce brushwood fire was built at the rock face to heat the stone. When it was as hot as possible cold water was suddenly poured over, the resulting drop in temperature fracturing the rock making it easier to tap.

Røros 'new' church

Bringing in the wood on a horse-drawn sledge.　　*AHC*

Trondheim (III) 1764–1766

A Selbumillstone at Selbusjøen. Feb. 2002.

RINGVE would be well known to the Bishop. Here the famous Norwegian Peter Wessel Tordenskjold was born. Now it houses a museum of music and the Botanic Garden (2000).

For more than a century the region and beyond was denuded of trees to feed the hungry demands of smelter and furnace and the avarice of owners, creating the most extensive treeless stony, sandy desert in Europe. The industry had seen rough, unhappy times as well as progress and changes in management had taken place. Miners' wages were low and mainly paid with special notes to be substituted for provisions, at an exorbitant rate of exchange, from Irgens' widow.

By 1701 Albert Angell (later the Grandfather of Karen Angell Suhm) of Trondheim had become the principal shareholder and the working of the mines left in the hands of a manager and engineers, architects and cashiers. Men of north German descent frequently held these important positions for the earliest engineers were Germans appointed by the king in Copenhagen.

In late July – early August 1734 Carl Linnaeus spent six days in Røros at the time Leonard Christian Borchgrevink was director of the mines. Borchgrevink was of German descent though his father had moved from Denmark to Trondheim in 1680 as cathedral organist. When he died in 1699 Leonard Christian was barely a year old. Around 1700 his mother, Catharina Borchgrevink, married Jens Finne, the town clerk at Røros, and so moved to the mining town with her young family. Jens Finne was a much-admired stepfather, the only father Leonard Christian Borchgrevink knew.

Linnaeus, still a student at Uppsala, had been staying at Falun, a copper-mining town in Dalecarlia, Sweden. The provincial governor had been so impressed by Linnaeus' survey of Lapland in 1732 that he offered to finance a similar survey of Dalecarlia. Seven fellow students from the medical department at the university begged to join him. The enthusiastic group left Falun on 3rd July 1734 and travelled more than five hundred miles by foot and boat, nightly recording in detail their observations. Røros was the furthest point reached, some forty miles over the Swedish-Norwegian border. This area proved of great interest to Linnaeus from the point of view of mineralogy for he had just travelled from such a site in Sweden. The group of naturalists returned to Falun on 18th August having spent forty-five days on the expedition.

When Gunnerus arrived at Røros on his visitation of Dalernes in 1764 many friends would receive him. Leonard Christian Borchgrevink and his wife Magdalene Brun had amongst their fourteen children one son who was particularly close to Gunnerus, Jens Finne Borchgrevink. He had been educated at Trondheim Cathedral School before entering the university in Copenhagen where he obtained his degree in 1758. During the last few months of his course there he had assisted Gunnerus before his appointment as Bishop and accompanied him to Trondheim in July. He was included in the bishop's entourage on this occasion to be warmly welcomed by his father and sisters who cared for his father since the death of his mother the previous decade.

(The name Jens Finne Borchgrevink is a little confusing. He was not actually related to his grandmother's second husband, the previously mentioned Jens Finne, but the close affinity of the family is stressed by this baptismal name.)

Peonies growing in the far north at Svinvik, Todalen (2000).

Trondheim (III) 1764–1766

Gunnerus collected a great many plants for his herbarium to be placed on his farm at Berg-i-Strinden (Berg prestegård) on the outskirts of the cathedral city. Particularly mentioned in Gunnerus' 'kopibok' of the time were the following specimen –

Centunculus minimus now classed as *Anagallis minima,* a member of the Primrose family. One of the smallest flowering plants (0.5–5cm) in northern Europe it is found on the sandy edges of the entire coast of the Scandinavian Peninsula. Its oval, stalkless leaves branch from the smooth stems with single, miniscule, globular pink flowers hidden in the base of the leaves. The English name CHAFFWEED suggests the shuffling noise the wind makes passing through the tiny plant, as through chaff or straw. The Norwegian name *PUSLEBLOM* indicates its minute size.

Campanula latifolia, the Giant Bellflower, Norwegian *STORKLOKKE (large bell)* has nettle-like leaves, sturdy angled stalks up to 130cm tall and distinctive purple-blue, occasionally white, flowers. The lower flowers open first. It stands boldly in clefts, crevices and fellside birch groves growing well in humus and natural leaf mold.

Alchemilla alpina, Alpine Lady's Mantle (FJELLMARIKÅPE) is found profusely throughout Norway in July and August. Usually 10–20cm in height, its pom-poms of tiny yellow-green florets above pretty deeply divided darker green leaves, are frequently hidden by more flamboyant coloured summer growth.

Henry Printz, in his article in the bicentenary booklet commemorating the birth of Johan Ernst Gunnerus, reported

Campanula latifolia
Giant bell flower
N. Storklokke

seeing these plants still growing in the same vicinity in 1918. Now modern plants of brick and stone occupy the immediate area in the spreading city although they may still be found in the wider environment of Trøndelag.

Short botanical forays were made in the immediate locality, for Gunnerus had established some plants in local clergymen's gardens, which he could easily visit without any interruption of clerical duties. The collection at Stene-i-Opstrinden,(now Steinan) Dr Henrici's home, was increasing and practical visits by the pupils of the Cathedral School were arranged here.

It is recorded that the bishop visited the tiny island of Tautra at one of the tips of the finger-like Frosta Peninsula, though it is Schøning's description of the monastery ruins that linger. Only one wall of the stone-built chapel remained and some stone foundations could be traced, for the thirteenth century monastery buildings had been of wood with roofs of turf. Abbey and monastery ruins survived in England because the roofs had been of lead over stone buildings. One can still admire the skill and patience of centuries of monastic workers in the decorative stonework. At Tautra one must stand in greater awe of the diligent, hard work of the Cistercian monks in the thirteenth century who cleared the land of stones to provide not only the building material for their church but also create pastures for sheep and fields for cultivation. These lands are still evident and bearing crops.

Although derelict from the sixteenth century, Schøning records in 1774 that the remains of a special herb garden

Alchemilla alpina
Alpine Lady's Mantle
N. *Fjellmarikåpe.*

Trondheim (III) 1764–1766

From Frosta looking across Trondheimsfjord.

Tautra Abbey remains (2000 AD).

Across Trondheimsfjorden from Frosta towards Trondheim.

were visited annually by the Trondheim apothecary for rare supplies for his pharmacy. Fruit trees, especially apple and morello cherry, were to be seen in what had been the monastery orchard a mile away, in the mid eighteenth century.

Sadness, disappointment and loneliness befell Bishop Gunnerus and Trondheim in 1765 when the Schøning and Suhm families left to reside in Copenhagen. They had shared the work of the Trondheim Society between them, specialising in historical and philosophical matters leaving scientific and theological articles to Gunnerus. Now devoid of stalwart companionship and in the midst of the preparation of 'Flora Norvegica' he must proceed alone.

Gerhard Schøning had been appointed professor of history at the renowned Sorø Academy in Denmark. Suhm had quarrelled so violently with Thomas Angell, who had even denied his own niece her rightful inheritance, followed his friend to Copenhagen in July 1765 taking his wife and only child, a four-year-old son, Ulrich Frederik, born like his mother in Trondheim, and an impressive library.

(This library later became the source of Suhm's income when he commenced a 'lending library' in Copenhagen. He angered some Danes when he published the names in a newspaper of those who did not return his books. Predeceased by his son in 1778 and first wife, Karen, in 1788, on his death in 1798 the books had been sold and eventually incorporated in the Royal Danish Library.
 The Royal Danish Library donated duplicate copies of Suhm's books to the Oslo University Library on its foundation in 1811. This university passed on further duplicated copies to the Royal Norwegian Scientific Society Library in Trondheim)

Gunnerus' concentrated work on his flora is apparent from his renewed correspondence with Linné. He took the opportunity to send a box of samples from his collection with two students going directly to Uppsala on 28th February 1766. Henrik Tonning was going to read medicine and Jens Finne Borchgrevink to be a private pupil of Linné for some months.

Sorø Academy, Denmark.

A further package was forwarded to Linné on 19th May including a complete example of the first part of his *Flora* and some pages of the second. Gunnerus expressed his profound gratitude to his kind friend who had patiently advised and corrected his work. He acknowledged his further indebtedness to Linné on the receipt of the information that he had been elected a member of the Royal Swedish Academy of Sciences.

The Bishop had intended visiting Nordland and Finnmark on his summer visitation that year but a storm swept ocean compelled him to return into the shelter of the Trondheimsfjord. There he made many local visits and was able to keep in close touch with the Trondheim Society, no doubt anxious about the forthcoming publication of the *Skrifter*.

During the six years of pastoral, observational tours Gunnerus realised the need for a systematic list of plants, which could be understood and used by the layman. He found many names for the same plant in different languages and dialects, many with apparently nothing in common. A great deal of time had to be spent consulting 'experts' who could not agree amongst themselves.

The red sign on the right-hand side of the road marks the road closure for winter. This road up to the Dovre plateau from Geiranger, is closed by snow four to six months each year. The only access to the village is by water (fjord, out to the sea).

Trondheim (III) 1764–1766

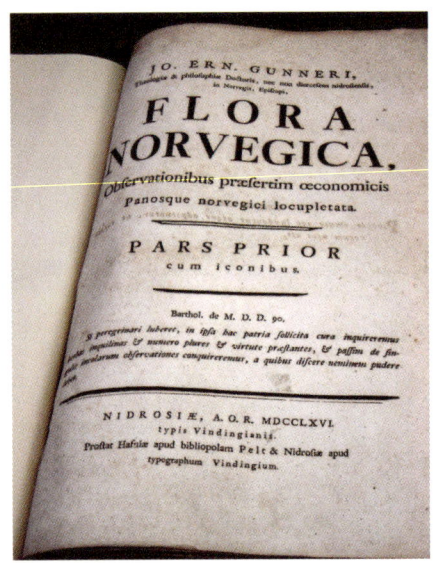

FLORA NORVEGICA
In safe keeping.

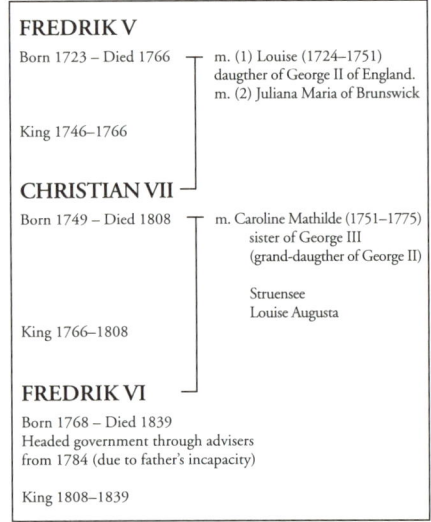

Royal Lineage

Linné had unconsciously commenced systematising plant names, using the universal language of the scientists of the day, Latin, and the application of the names of the reproductive parts to group the plants in his own garden. Frequently he added a second name for his own amusement and to help his memory, a name reminding him of some one or some association. He did not take the binary nomenclature seriously nor did he consider 'Systema naturae' his main work, although it had reached the tenth edition when Gunnerus became bishop of Trondheim.

This memorable year of 1766 saw the publication of the first part of, *FLORA NORVEGICA* by Johan Ernst Gunnerus, dedicated to the King. As the book was being prepared it would be presumed this would be Frederik V, the official instigator of Oeder's work, but he had died this year and been succeeded by his son, Christian VII. In the introduction Gunnerus paid great tribute to the clergy who had helped him throughout his ministerial visitations. He realised there was much work still to be done on the publication, parts would need correcting as he gleaned more information but a start had to be made.

Three hundred and fourteen plants were listed each named in Latin, using as far as possible the name recognised in Linnaeus' system, and Norwegian, often in several forms. Many plants also had Danish, German, French and English names. Economic and medicinal uses were given. Gunnerus hoped the details would be extended as the priests acquired more information through the dissemination of available material.

Early November Gunnerus took the opportunity to send some specimen to Linné with Bishop Friis who was on his way to Copenhagen. The three items were a gift for the Natural Science Academy of Stockholm and would be explained by Tonning. Towards the end of the short letter, dated 8 November 1766, accompanying the package mention is made for the first time of *Gunnera*. No record is available of what had transpired in the intervening months. How had Gunnerus received such startling, praiseworthy news from Linné that a plant family was to be named in his honour? He expressed his thanks and stressed how he could not have achieved such a distinction without Linné's assistance.

Carl Linnaeus (1707–1778)
Portrait in oils by L. Pasch after A. Roslin,
1775, copied for Sir Joseph Banks, now in
LINNEAN SOCIETY OF LONDON

7

Trondheim (IV)
1767–1771

Pastoral Visitations

The following year Gunnerus embarked on his third northern visitation. Two personal assistants, Jacob von der Lippe Parelius, who was later to take charge of the Scientific Society's collection and Jens Finne Borchgrevink, accompanied him. The bishop embarked with high hopes and expectations for not only had his own personal scientific knowledge improved but also he had two young experienced associates in his party. Borchgrevink had lived and studied with Linné for nine months the previous year on his estate at Hammarby, a few miles from Uppsala. Here the garden and farm were well organised and Borchgrevink must have frequently taken the upward path through the woodland to attend the master in his 'cabinet' – the study Linnaeus had built to contain his collections and where he could lecture.

Trondheim (IV) 1767–1771

Model of lecture stool, which Linné invented for use when lecturing in his 'cabinet'.

The decoration on the cup and plate of Swedish porcelain shows the delicate plant Linnaea borealis which the great botanist named after himself.

*Linnea borealis L.
From Flora Danica*

Bishop Gunnerus would now be a familiar visitor in the diocese of Helgeland, Salten, Lofoten and Troms but this year he was determined to sail round Magerøy,(the North Cape) and on to Kjelsvik in Finnmark, for the sake of his two assistants. The Scientific Society's collection of shells, minerals and plants was enlarged.

Some new specimen were found and one in particular, *Gentiana serrata* Gunn, was listed in Flora Norvegica 819, as having been first identified by Gunnerus, believing it a new sub-species. However it was later found to be *Gentianella detonsa* Rottb. *(FJÆRESØTE)* ; It is still only rarely found along the northwest coast of Norway

A Royal Decree was issued on 17th July 1767 that the Society of Trondheim should henceforward be known as *the Royal Norwegian Society of Sciences and Letters (DET KONGELIGE NORSKE VIDENSKABERS SELSKAB).* The Crown Prince was named President and Johan Ernst Gunnerus vice-president. It was an honour that a Norwegian Society should be named in this manner. The question was raised, was this a step towards the foundation of a Norwegian university in Trondheim? This had been the dream of Gunnerus and his colleagues from the days of group discussions in Suhm's library and hinted at in the bishop's letters. Hope was revived.

When the King's birthday was celebrated on 29th January 1768 at the Royal Norwegian Society of Sciences and Letters the vice-president addressed the meeting. Gunnerus mentioned it was his dearest wish that the acknowledgement of the Society would lead to their own Norwegian university.

The mid-summer visitation was made along the southern shore of Trondheimsfjorden, the bishop's ship being used as study/ base, for most parish churches were accessible by boat. It must have been a colourful sight to see the boats of various sizes being rowed across the fjord, bringing joyful families to celebrate the entry into adulthood of two or three confirmation candidates. The church would have been built on the higher ground of a promontory so the boats could be tied or beached round the base. The grassy slope, cropped by a few tethered goats, and a few silver birch trees, covered by this time in foliage – small delicate, shimmering platelets like a fountain of palest green, provided the perfect gathering place. As the sound of the church bell echoed across the fjord the community understood the clergy had arrived and so were drawn together.

Romsdal and Nordmøre

There would be feasting and no end of music (at least one fiddler) and jollity, as old and young congregated after the service on the hillside – Holy Days became holi-days! Gunnerus recorded visits to Aure and Tingvoll, ancient sites of worship, possibly dating to heathen times. The name Tingvoll refers to 'the meadow where the Ting (free men) assembled' in the Middle Ages and the stone church dating from circa 1150 is one of the best preserved in the whole country.

Grytten, where the River Rauma flowed into the Romsdalsfjord, after hurriedly slashing its way from Dovre and Trollstigen in gigantic steps to the sea, is no longer found on the map. The river valley now also carries rail and road links from the main north-south highways and the busy junction of Åndalsnes has developed on the site.

A typical church setting on a spit of land in a sheltered fjord. Ullensvang church, on Hardanger fjord has been a prominent site for centuries.

Trondheim (IV) 1767–1771

Grytten has developed into Åndalsnes. Rowan trees on right.   *AHC*

Rowan Tree at Moreton Morrell in autumn with its colourful berries.

Correspondence between Gunnerus and Linné was now more regular, frequent and scientifically detailed. There were frequent reference to Dr Tonning and Hr Borchgrevink and of particular interest in Gunnerus' letter to Linné, written 20 October 1768, is mention of the fact that Tonning was with him in Trondheim prior to travelling to Stockholm and Copenhagen but Borchgrevink was accompanying Professor Maximilian Hell from the Vienna Observatory to Vardø and would be away until the following autumn. The scientific object of this journey was to observe the transit of Venus, on 3rd June 1769, from the most northerly point of Europe. At the same time an expedition with a similar purpose was leaving Plymouth, England. Capt Cook and Sir Joseph Banks were on board *the Endeavour* bound for Tahiti.

As the bishop continued his pastoral work, visiting as many areas of his diocese as possible, his interest as a botanist was enhanced by many personal encounters with clergy and local residents. His main visitation this year (1769) was spent in the deanery of Innherad which stretched from the inner Trondheimsfjord and the well-worn main track through Stjørdalen and Teveldalen to Sweden in the south, northwards along the Swedish mountains to the hinterland of Namdal at its northern extremity and bounded on the west by a botanist's paradise, Lake Snåsa and the head of Trondheimsfjorden.

In the surrounding limestone area of Snåsa Gunnerus found several rare orchids –

Epipactis palustris, Marsh helleborine (MYRFLANGRE) with up to fifteen florets of violet-brown outer petals and two inner petals of crimson-white and an enlarged lip of crimson and white stripes surrounded by a white frilly edge growing to a height of between 15 and 30 cm. It is rarely found in Norway today.

Neottia nidus-avis, is vividly described by its name – the Birdsnest orchid, FUGLEREDE, for the habitat of a beech forest mat of leaves conceals a flower of the same colour, honey-brown, and more easily distinguished by its sweet, sickly fragrance.

Ophrys (myodas) insectifera, FLUEBLOM, is one of the rare insect-like orchids, with apparently black backed insects crawling up the stem as flowers.

Gunnerus particularly noted *Cypripedium calceolus* MARISKO. It later appeared in Part II of 'Flora Norvegica' as specimen MCXIII (1113) on page 147.

Epipactis palustris
N. *Myrflangre*
E. *Marsh helleborine*

> I have seen varieties of Cypripedium in orchid exhibitions and in our own shade house in Singapore but they were inconsequential after 16th June 1999 when my complete attention was grasped by three wild Cypripedium calceolus standing like upright sentinels at the entrance to their secret lair. Time stood still. The narrowed river rushed over its rocky limestone bed, fringed with willow and birch, unheard. A picture, etched, I hope forever, in my mind. There were hundreds more in the area, north of Snåsa but I will not reveal too many secrets for those who truly love such God given treasures will surely find them.

The striking, erect plant, *Cypripedium calceolus,* Lady's slipper (*Marisko*) grows to a height of 40cm on a sturdy, soft downy,

Cypripedium calceolus
N. *Marisko*
E. *Lady's slipper*

Trondheim (IV) 1767–1771

Cypripedium calceolus L.

The birth of a book, November 2001, on the site of Gunnerus' home. Left to Right, standing Stein Johansen, Ingar Lomheim Seated, Harald Nissen, Monica Aase.

green stem round which the oval-oblong-pointed, deeply-ribbed veined, bright green leaves are sheathed. The stalk usually bears one flower, only very occasionally two, with a strong, robust lip, more suggestive of a favourite ballet-shoe toe than the delicate lady's slipper of S.E.Asia. The bright-golden yellow of the lip is red spotted inside and the dark red-brown petals protectingly bow over it. The whole is camouflaged by its light woodland habitat where, until closer inspection is made, a patch of the orchids appears like dappled sunlight through the trees, the cup a touch of clear sunshine.

Gunnerus had bought various pieces of land and property during his time in Trondheim but he sold all in the year 1769 in order to purchase, on 27 October, a property on the corner of Dronningens gate and Apotekerveiten, then named 'Schultgaarden'. The Royal Norwegian Society's library was housed here for it was in a convenient situation in the town and was open to all students on Wednesday and Saturday afternoons. (It had previously been in the Cathedral School).

The Norwegian word *gård -en* is defined as a farm or a manor house but from usage I found it referred to any detached house which had land all the way round. Norwegian friends named my small detached bungalow in Yorkshire, *Myrgården* (the Marsh farm, for my maiden name was Marsh) but a less likely 'farm' I could not imagine. No doubt the word 'garden' comes from this Norse root, as does 'yard'. The Norwegians call a garden- *hage;* a garden nursery-*planteskole;* a market garden- *handelsgartner;* and a botanical garden- *botanisk hage.*

The Britannia Hotel has occupied the site of 'Schultgaarden' since 1896. It was in the Palm Court of the Britannia Hotel that I had my first meal in Trondheim in 1960 and it was there too that this writing first began to take a tangible form in November 2001.

Professor Peter Maximilian Hell and Jens Finne Borchgrevink visited Gunnerus in Trondheim en route for Copenhagen on their return from Vardø in September 1769. Frequent references are made to the travellers in the Gunnerus–Linné correspondence during this year and the following. Gunnerus was still at work on 'Flora Norvegica' and had received further details and information from Prof. Hell which he shared with Linné.

By 1770 the Bishop of Trondheim had developed a special affinity for Northern Norway and so on his summer pastoral visitation he again journeyed as far north as Tromsø. Amongst the plants he gathered for his collection was one that is officially attributed to Gunnerus.

Arenaria norvegica Gunnerus
The tiny Arctic sandwort *(SKREDARVE)* in loose tufts merely 3 to 7 cm high is found mainly in northern Norway in Fennoscandia and abundantly in Iceland. It straggles along bare limestone screes from a matted root system, its dark green fleshy oval leaves, in pairs with a trace of a central vein, only 3 to 4mm in length. Each stalk produces one or two flowers of five white petals with a hint of pink and yellow pollen bearing anthers. It flowers mainly between June and July but it can be found later.

With the help of Editor, Stein Johansen, Håkon A. Andersen, curator of the National Museum of Decorative Arts, carefully unwraps the heavy wrought iron segments of the balustrade made for Bishop Gunnerus' home in Dronningensgate.

The railings were placed on either side of the steps from the road side to the main entrance, the left hand side one has the word ANNO (year) and the right 1770 beautifully wrought and interwined with patterned oak leaves.

Trondheim (IV) 1767–1771

Arenaria novegica Gun.
N. Skredarve
E. Arctic sandwort

Even in the midst of a rich business and administrative community, and the activities of the museum and '*skrifter*', there were periods of loneliness when Gunnerus longed for the deep companionship of like-minded friends, as there had been when he first arrived in Trondheim. Schøning and Suhm kept in contact but it was not the same as chatting round a library table in the warmth of lamp lit friendship. His pen-friendship with Linné continued, indicating that his natural scientific work was progressing, if less enthusiastically.

Twilight on the fjord.

8

Copenhagen (II)
1771–1772

Struensee

The bishop's visitation in 1771 was in Dalernes, the rich area south east of Trondheim. He had been centred for some days at Støren in Gauldalen, when he received a summons from Struensee, the Chief Minister of Denmark, to go immediately to Copenhagen. Gunnerus' advice was needed in the restructuring of Copenhagen University.

At last! The opportunity for Norway, particularly Trondheim, was opening. Hope was revived and enthusiasm rekindled. Without a moment's hesitation, on 20 July, he cut short his visitation and turned to botanical investigations in Gudbrandsdal on his way, by coach, to Copenhagen.

One cannot imagine a more uplifting, gentle journey southwards, down the great Gudbrandsdalen on a mid July day. Once over the dreary Dovre plateau the frothy icy blue-green

Copenhagen (II) 1771–1772

Trondheim to Copenhagen via Gudbrandsdalen.

waters of the river Lågen begin to spread out and the valley broadens into centuries old cultivated farms worked by the same families for generations. The groups of seter huts could be spotted high up the mountainous sides and they would have been lively, as the bishop hurried down the valley with his face set on Copenhagen that summertime. He did not miss the wayside beauties of nature and added to his collection of specimen.

As hospitality was extend in parsonage or manor house he would take the opportunity to discuss the situation in Copenhagen and possibilities in Norway – and the natural history of the area, in the long, bright, summer evenings.

The stately command and the anticipation of future organisation must have somewhat prevented the botanist from tarrying as he would have desired round Norway's largest lake, the glorious Mjøsa or amongst the ruins of the old cathedral in Hamar. On 23rd August (1771) he reached Fredrikshald (now Halden) and crossed out of Norway, along the Swedish west coast, eventually reaching Copenhagen in October.

University

Excitedly Gunnerus sent a short letter to Linné informing him of his arrival in Copenhagen on 3rd October. He had already visited the King and Queen and taken a quick look at some copperplates being prepared for the second part of Flora Norvegica. He was hoping to see the diary and any surviving part of the collection of Professor Peter Forsskål of the Arabian Expedition, from which only Carsten Niebuhr had

Sheep on Dovre plateau in summer.

recently returned. Forsskål was a particularly favourite 'disciple' of Linné and had fortunately managed to send some details directly to him from the unfortunate expedition.

Communications must have been considerably speedier between the centres of learning – Uppsala and Copenhagen – than Uppsala and Trondheim, for Linné's reply was penned on 26th October. He was disappointed Gunnerus could not include Uppsala on his journey to Copenhagen but he stressed that *Gunnera* must be visited in the Copenhagen botanical nursery where it must have grown in greatness, like the Bishop.

By mid-December Gunnerus had submitted his suggestions for the reorganisation of Copenhagen University to include a Norwegian university. Some interest was shown but the new university could not possibly be in Trondheim for that was a place too far away to be used by Danish students. It could possibly be situated in Kristiansand for that would be a simple sea journey from Jutland, the portion of Denmark adjoining Europe. Gunnerus was disappointed and began to realise that Norway could only have its own educational establishments if it was separated from the kingdom of Denmark. This was the germ of the idea Schøning, inspired by his predecessor at the Trondheim Cathedral School, had been quietly working at for two decades in his writings.

I wonder if the three old friends Schøning, Suhm and Gunnerus were able to meet in those treacherous days of January 1772. Copenhagen was a city fermenting with discontent. On the death of Frederik V in 1766 the seventeen-

Ringebu Church, Gudbrandsdalen.

Gudbrandsdal looking towards Lake Mjøsa.

Copenhagen (II) 1771–1772

Spring comes to Bregentved.

Gudbrandsdal festbunad, an example of the author's own handwork.

year-old crown prince was proclaimed King Christian VII. He had a neglected, unfortunate, early upbringing and was considered by some as having an unsound mind. His marriage was immediately arranged with the fifteen-year-old granddaughter of King George II of England, Caroline Mathilde, an heir was needed and it was believed the marriage would be a stabilising influence.

The wealthy, influential German class continued their domination, as in the previous reign, until a clever, young physician who had accompanied the king on a European tour, 1768–69, was appointed Court Physician and then cabinet secretary. The thirty-year-old Johann Friedrich Struensee had complete power over the schizophrenic young king, and younger queen, after reportedly saving the life of their child, the crown prince. His ambition knew no bounds.

As Count Struensee, he wielded complete power from March 1771 until January 1772. He introduced many reforms, such as freedom of the press and a reduction of compulsory peasant services. He wished to restructure the university and knowing of Gunnerus' abilities in organisation had sent for him in July 1771 for consultation.

Struensee's reforms, without time for opposition, alienated many officials and he was the victim of a conspiracy, during a court ball, in January 1772. He was arrested and later publicly brutally executed, the accusation being his liaison with Queen Caroline Mathilde. (They did have a baby daughter). Her brother, then George III of England, permitted her to remain, in solitary confinement in a family property in Celle,

Hanover, the last few years of her brief, sad life, so saving the queen from a similar fate.

Oeder had been dismissed as the editor of 'Flora Danica' after the fall of Struensee but was later appointed Land Marshal in Oldenburg, a paid government post. Gunnerus longed to return to Trondheim and leave the turmoil behind. He wrote to Linné on 29 February 1772 of his desire to be away from the capital but there were still some in authority who might help. The few who showed sympathy for his mission were overwhelmed by the many afraid of the diminution of Danish power.

The homeward journey commenced at the beginning of June and Trondheim was thankfully reached on 1st July. A very disappointed bishop wrote to Suhm, still living in Copenhagen, "I am heartily glad to be home and never wish to see Copenhagen again".

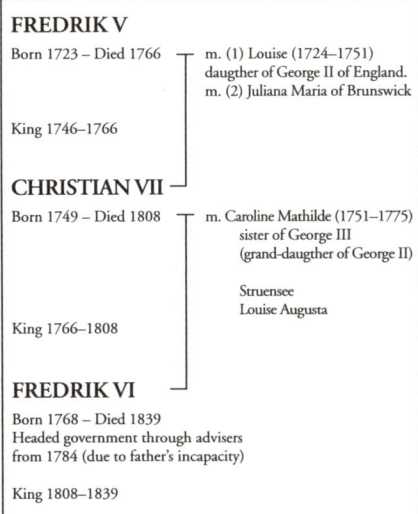

Royal Lineage

Vikings !   *AHC*

Copenhagen (II) 1771–1772

'The Lord gave us memories that we might have roses in December'.

Growing up the wall of Rosendal Manor House, sheltering below a snow-covered mountain, roses were growing in December 2001.

Trondheim (V)
1772–1773

Bishop Gunnerus resumed his pastoral work by 3rd August 1772 and set out to continue his interrupted visit to Dalernes *prosti* of the previous year. He found solace in the wonders of God's world at his feet, collecting more than one hundred different plants to include in his herbarium, as he returned up Gauldalen to Haltdalen and on to Røros.

There would be joy and sadness on the bishop's arrival at this time. He would be interested in all the welfare work provided for the manual workers and in the school beside the church. Members of the Borchgrevink family were occupied in the copper works, church and town but stalwart, Leonard Christian, twenty years older than Gunnerus, was less active now.

Gunnerus had returned to Trondheim by 5th September having visited Oppdal at the head of Sunndalen, and the valley which led down to Surnadalsfjord, a branch of Romsdalsfjord,

Trondheim (V) 1772–1773

Return to Dalernes.

Åndalsnes (Grytten)   AHC

and had continued north, round the fabled Trollheim massif (*the home of the legendary trolls*) and so down Orkdalen to the parish of Orkanger. Perhaps Gunnerus was so weary of land travel that he completed his journey by ship – it would be a short, comfortable, relaxing journey within the sheltered fjord, home to Trondheim.

Niels Dorph Gunnerus, the orphaned nephew who had lived with the bishop's family in Trondheim, from 1763, and attended the Cathedral School, found his dear uncle no longer amusing and cheerful but quiet and morose. As the long dark days of winter were passing (March 1773) Gunnerus wrote to Suhm in Copenhagen and spoke of his isolation. He was considering a change, moving closer to his peers. He had heard that Bishop Nannestad would like to retire from Christiania and he considered moving to his old hometown, taking the organisation of the Royal Norwegian Society with him. Christiania was a more convenient focal point for the world.

Then in June came the heartening news that the king had granted a substantial monetary award to Gerhard Schøning, to be used for historical research during his forth-coming sabbatical year, which would be spent in Norway. The prospect of his friend being in Trondheim was a tonic and arrangements were made for the two old friends to tour the well-loved *prosti* of Romsdal and Nordmøre, combining work and pleasure.

Unfortunately the annual visitation of the bishop had to be delayed from the customary summertime that year. As vice-president of the Royal Norwegian Society of Sciences and

Letters, Bishop Gunnerus had to formally receive Prince Carl of Hessen and his guests, on the official visit to the society. The royal party arrived in early August.

It is not quite so easy to trace the register of the bishop's tours from 1771, for it is recorded in a different format. From this time entries were under the various Parish names, rather than in complete chronological order. The first entry for 1773 was at Hemnefjord on 12 August. They had sailed round the Agdenes Peninsula in the Trondheimsfjord, west of the city, into the Hemnefjord towards Kyrkesæterøra – a name reminiscent of the Swedish church villages at the mouths of great rivers on the Gulf of Bothnia (Luleå, Umeå) or summer pastures belonging to the church.

The entry itself is in Gunnerus' writing and many names are mentioned. A list is given of the peripatetic teachers, who moved from farm to farm for a few weeks at a time in the parish and the number of pupils they taught. This would be done particularly for the benefit of Schøning, as the teachers would have been his ex-students from the Cathedral School.

Returning to the main fjord some days were spent sailing round the islands until on 15 August confirmations were recorded at Aure. Three days later, on 18 August, Herr Borchgrevink preached at the church in Surnadalen. The writing of this entry is blotched and almost illegible. Was the bishop unwell?

A pause was made at Tingvoll on 22nd August and at Sunndalen on 24th before reaching Nessset on Romsdalsfjord on

The river Rauma approaching the Romsdalsfjord. Two farms can be seen in the fertile bend of the river. (Mjelva).
AHC

Geiranger showing the typical route of melting snow from the mountains, cascading in a powerful torrent down into a fjord, creating a pocket of fertile land at the head of the fjord.

Trondheim (V) 1772–1773

River Rauma dashing down Romsdal.

Romsdalsfjord near Molde.   AHC

27th. There the bishop preached on the text from St Paul's second letter to the Corinthians, chapter five and the twentieth verse,

> 'We are therefore Christ's ambassadors, as though God were making his appeal through us. We implore you on Christ's behalf: Be reconciled to God.'

It should have been a pleasant sail down the protected Langfjord to the ancient church on the tiny island of Veøy, another ancient gathering place, where they would have joined the main Romsdalsfjord. A week later the bishop's party reached Molde and there Hans Strøm preached. It was a short journey down the Romsdalsfjord to have taken a week but by 5th September (1773), daylight would be shorter, darkness hampering sailing and autumn storms gathering, all delaying factors.

Hans Strøm would be a very welcome addition to the group. He was born, one of twin boys, on 25th January 1726, in Borgund, Sunnmøre, where his father was vicar (*sogneprest*). Sunnmøre was in the adjacent diocese of Bergen so the boys had attended Bergen Cathedral School before going to Copenhagen University. Hans was a theologian, a keen naturalist and mapmaker, whilst his brother was a lawyer. He was a great admirer of Erik Pontoppidan, who had been the Bishop of Bergen for eight years (see page 62) and a personal friend of Gunnerus, for they had many interests in common.

In one of Gunnerus' frequent letters to Strøm, who was a member of the Royal Norwegian Society, he had expressed the wish that they might live nearer each other so they could discuss and correspond quicker.

> (Strøm's uncle, his mother's brother, Eiler Hagerup had been the Bishop of Trondheim 1731–1743.)

From 7th September 1773 the entries in the second volume of 'Trondheim Biskop Visitasprotokol' no longer appear in Gunnerus' writing. Turning their backs on the many peaks of the Romsdal Alps the party sailed northwest through Julsundet to the ancient fishing port of Bud, on the outer coastline bordering the North Sea. Two centuries before Bud had been the most prosperous place in Romsdal – a 'National Assembly' was held there in 1533 – now just a tiny, colourful fishing village with the aura of history, facing the uncompromising sea.

Along the coast from Bud northwards to Kristiansund is Hustadvika, the most vicious stretch of water to be negotiated approaching Trondheimsfjord.

> (Some forty years ago it was the custom to disembark northbound cruise passengers from the Express Coast Steamer at Molde and transport them by motor coach overland to Kristiansund to avoid this rough passage. Today such cruise passengers still leave the modern vessel at Molde, the City of Roses, to experience a journey on the Atlantic Highway, the road which climbs and drops, joining islands and is more beautifully exciting than any fairground thrill.)

Unfortunately Hustadvika was particularly rough in early September 1773 and our dear bishop was very seasick. The ship sought the shelter of the islands at the first opportunity, turning into Kvernesfjord and anchoring at Kvernes, on the

Romsdal Alps

Romsdalsfjord approaching the sea.

The fishing village of Bud today, sheltering in a creek from the North Sea. It was an ancient gathering place.

Trondheim (V) 1772–1773

Across Kvernesfjord. The new Atlantic Highway in the distance.

The spit of land where the parsonage stood in the background. Here Gunnerus rested five days on his last journey.

southern side of the island of Averøy. There Gunnerus rested some days in the parsonage, attended by anxious friends but too far away from the ministrations of Dr Henrici.

There are two churches, prominently high on the peninsula, sheltered from the North Sea blasts, yet providing another convenient meeting place for fjord traffic from c. 1300. Side by side stand the earlier stave church, built entirely of wood, which had a chequered history, and the 'modern' nineteenth century, white-painted building. The parsonage, containing all the church records had been destroyed by fire in the early nineteenth century and no trace whatever remained of Bishop Gunnerus' last visit.

Hr Strøm took charge at Kvernes and in the register, to be seen in the Trondheim Statsarkivet, dated 12th September, he referred to the Bishop's 'inner sickness'. After some days bedfast, Gunnerus' fever had abated and on 15th September he was able to continue to Kristiansund. The illness did not ease, signifying it was more than seasickness and Strøm's comment suggests a stomach disorder – many, many complaints were undiagnosed nearly three centuries ago.

In the early morning of Saturday 25th September 1773, in his fifty sixth year, Johan Ernst Gunnerus died peacefully with Schøning and two more close friends at his bedside.

The body was conveyed to Trondheim where it was received with great sorrow and laid in the cathedral crypt on 18th October 1773. Schøning addressed the memorial service in the Cathedral School on 1st March 1774.

(Bishop Gunnerus' remains were reburied in the graveyard of the Domkirke in 1867 when extensive renovations were being made in the crypt.)

I did not want to come to this part of the account. My feelings were with Sister Stine, and the friends who must take up the threads of Gunnerus' many activities, but there is no mention of them.

Then I realised … it was not an end but a beginning! Attempt to destroy a globule of mercury and it becomes many more, each indestructible. So the spirit and influence of each indestructible faithful soul lives on as *Gunnera* does.

Kvernes old stave church.

Trondheim (V) 1772–1773

Trondheim *Dreier c 1800*

10

GUNNERA

There are many names of plant families in the botanical world actually bestowed by Linnaeus, which envelop biographical details of some student, (sometimes referred to as 'apostle'), official or friend. Will these vanish in the future when plants become classified by DNA and appear as letters, symbols and numbers? Through his influence, Linnaeus was able to send many disciples around the world, as chaplains or doctors, aboard the vessels of the Swedish East India Company. The Reverend Olof Torén was the only one to sail round the Malay Peninsula to Indo-China and amongst the collection of seeds be brought back to the master, from two tours of duty between 1748–49 and 1750–52, were some Linnaeus named *Torenia,* perpetuating his name. *Torenia asiatica,* in various forms, is still to be found included in the ground cover in some Malaysian oil and rubber plantations.

It was reading of the controversy generated by the naming of *Forsskalea* that prompted this search for *Gunnera.* Peter

Gunnera

Gunnera sp., Mount Tomah, Botanic Garden, Sydney, NWS, Australia. (early spring).

Royal Botanic Garden, Copenhagen. Gunnera tinctoria on right.

Forsskål (born 1732) was a clever Swedish student, who had studied at Uppsala and Göttingen, and was the botanist on the ill-fated Danish expedition to Arabia, which departed from Copenhagen in 1761.

The only member of that expedition to return was the admirable, undaunted Carsten Niebuhr. When Linnaeus named a plant, raised from seeds forwarded by Forsskål, *Forsskalea*, Niebuhr was horrified that a stinging nettle should be named in honour of the natural leader of the group. On careful consideration of the specimen, Linnaeus had indeed made a suitable choice, for Forsskål himself had described it as stubborn, wild, obstinate and angular – the characteristics that had carried that expedition forward.

Why *Gunnera* in honour of a Norwegian bishop?

It is recorded that the original seeds from which *Gunnera* had been developed, was from a southern hemisphere expedition, delivered to Copenhagen or Leyden in Holland. Unfortunately many of the strange samples brought in the early eighteenth century were jealously guarded by some or neglected by others and details misplaced. Providentially some came into the hands of caring scientists like Linnaeus and his students in Uppsala, Oeder in the Royal Botanic Gardens in Copenhagen and Joseph Banks in England, who had taken Solander, another Swedish student, on Captain Cook's expedition to Australia.

On 1st May 2001, growing on the roadside by Bregentved, the estate officially given to Moltke, Frederik V's closest friend and patron of the young Gunnerus, were to be seen masses of strong, sturdy 'butterbur' (*Petasites*) growing 40cm above dark green, heart-shaped leaves. The masses of tightly packed pinkish-white flower heads, protected by beige, strap-shaped bracts, were reminiscent of the developing crown of flowers forming the base of *Gunnera manicata* in the Royal Botanic Garden, Copenhagen. Some days later Peter Wagner, in charge of the treasure store of the Botanical Library, Copenhagen and Oeder specialist, informed me,

"The plant to be renamed *GUNNERA* by Linnaeus was brought to the botanical garden in Leyden. Paul Herman first described it in his catalogue of the said garden under the name of *Petasites africanus*. Herman tells it was brought there from the Cape of Good Hope."

The name *Petasites* comes from the Greek word for a broad brimmed hat, providing shade for field workers in the noonday heat. (W.T.Stearn)

Herman (1646–1695) was a German botanist physician, professor at Leyden in the Netherlands who had spent some years in Ceylon and so must have travelled round the Cape of Good Hope. It was the only route, for the way across Asia Minor was unknown and the Suez Canal unimagined. Linnaeus visited the renowned garden at Leyden several times between 1735 and 1737, whilst completing his medical degree at nearby Harderwijk University.

Gunnerus had written to Linnaeus from Copenhagen, in April 1772, on whose advice he had looked for *Gunnera* in

Detail of Petasites from roadside photograph below.

Bregentved Farm, Denmark. Roadside 1st. May 2001.

Details of developing Gunnera sp.

Developing buds of Gunnera tinctoria, Mount Tomah, Botanic Garden, Sydney, NWS, Australia.

the botanic garden, but had seen only the flower spike. Professor Rottbøll had insisted on showing him other specimen and the place in another garden where *Gunnera* had been held previously, before Gunnerus was able to view the flower. He was disappointed not to see any foliage but this was because of the season.

The flowers emerge directly from a rooted clump in the ground (rhizome-like) in late spring and with the lengthening daylight of summer (June in Scandinavia, December in Australia) the leaves unfold from a basal, fibrous, coconut-like ovoid, into enormous leaves on stout stems each often 2m long. In late summer (September in Scotland, March in Australia) the whole plant appears like a gigantic crown of rhubarb, *Rheum rhaponticum,* creating a massive shady chamber to protect the now fruiting flower. The thick, green-red, leaf-stalks are coated with bristles if not thorns. It is not surprising that in later years the plant has been nicknamed 'apes' rhubarb', but one should not attempt to eat it, like the delicious rhubarb grown in the acid soil of coal-mining areas of Northern England and so useful for setting jams and filling pies.

> (Nor is it related to Rheum officinale, the native of Central Asia, whose bitter, astringent roots were dried and used for medicinal purposes. In my childhood a small bottle of 'tincture of rhubarb' or 'Turkey rhubarb' was always available after feasts – one recovered with the threat of it!)

The genus is confined to the southern hemisphere and is especially found in parts of South America, as is indicated by the nomenclature of some of the largest species, for example

– *G. brasiliensis* (Brazil) = *G. manicata*, *G. chilensis* (Chile) = *G. tinctoria* . It is a perennial which although it is naturally self-seeding the tiny fruits are prevented from travel by the hefty stalks and leaves which die back in winter and decay, protecting and nourishing the crown.

In contrast to the powerful height and breadth of the two sample varieties named and seen in splendour and photographed in Mount Tomah, in the Blue Mountains of New South Wales, Australia and Inverewe Gardens, Scotland, there are tiny, delicate varieties whose flowers are difficult to see with the naked eye. They creep along the ground forming a bejewelled, emerald and amethyst-tinted cover. It was an exciting, enthralling sight to see *G. magellanica* carpeting the brow of the steep hillside above the Todalsfjord, in Svinvik Botanic Garden, within the diocese of Trondheim, Norway. (The *G. tinctoria* was not happy in the climatic conditions there.)

Details of Gunnera chilensis in Copenhagen, late summer 2001.

Gunnera magellanica, Svinvik Arboretum, NTNU. NB! The tiny hand-held flower.

Gunnera

Gunnera tinctoria,
Inverewe Gardens, Scotland, early autumn

Gunnera manicata,
Inverewe Gardens, Scotland, early autumn

The small, kidney shaped leaves with a scalloped edge were a bright dark green, edged with purple and purple veins beneath. Specks of light, through the pine trees, flickered on their shiny surface. The 5–7 cm wide leaves on 6–10 cm long, fragile stalks had to be gently parted to reveal the concealed miniscule, pink-mauve, cone shaped panicle about 1 cm long. As the name, *G. magellanica* would suggest this plant was found in profusion around the Strait of Magellan, between Southern Chile and Tierra del Fuego, discovered in 1520 by the Portuguese navigator in the service of Spain, Ferdinand Magellan, attempting to reach the East Indies by sailing westward across the Atlantic Ocean.

Today the genus *Gunnera,* including some forty species, is included within the family *Haloragaceae,* the water milford family of dicotyledons. Large or small, *Gunnera* thrives beside water.

As we saw the insight and wisdom of Linnaeus in the naming of Forsskalea, we again perceive the depth of his understanding of human psychology. From the 'broad brimmed hat' to the extensive shade provided by the broad leaves of *Gunnera manicata* we have the indication of a broad-minded man of wide interests and deep knowledge.

Johan Ernst Gunnerus had developed from a mini-herbalist, and fluent classicist, in Christiania, through studies, teaching and writing in philosophy, natural science, mathematics and oriental languages (Hebrew and Greek) in Copenhagen and North Germany, to theology. This subject bound his talents and ever developing interest in the world around him together, into the mature Gunnerus.

11

FLORA NORVEGICA
From 1766

Determination to visit Trondheim during its Millennium Jubilee year found me making enquiries for Gunnerus' 'Flora Norvegica' in December 1997. I had sympathetic advice from Mox-næss booksellers and it was only then I realised I was looking for something rather special. As suggested, I visited the antiquarian bookseller, where on climbing the stairs I was regarded with disdain and my enquiry rejected as futile.

The following week I was making my way towards Karl Johan's Gate in Oslo and I was attracted to the clean, orderly, brightly displayed window of an antiquarian bookseller. I crossed the street to enjoy a closer inspection of the artistry of the bookseller. I had given up hope of 'Flora Norvegica'…but I could ask. Mr 'Norli' was charming, the catalogue was produced and an original copy of 'Flora Norvegica' (minus one plate) was available. The deal was made.

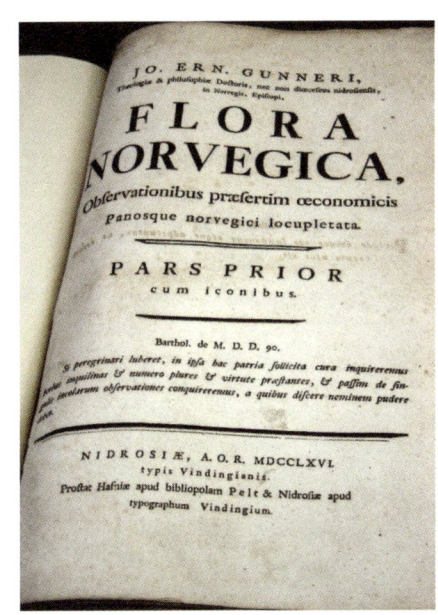

The FIRST PAGE of the Warwickshire FLORA NORVEGICA.

Flora Norvegica

The River Nid, Trondheim. July 2000

On display in oak box, carved orchids (as from Singapore) on each corner-

and the mouse trade mark beneath.

I did not see my 'Flora Norvegica' for more than a year for I had it delivered into the safe keeping of friends who would nurture it in the manner to which it was accustomed. I was living in the tropics and knew that in the constant heat and humidity of my home it would deteriorate, may even disintegrate. The opportunity to take possession of the book did not occur until June 1999 by which time I had decided I would present it to the Warwickshire College, England, where it would be deposited in the Looi Horticultural Library on the Moreton Morrell Campus.

To be displayed effectively, the famous 'Robert Thompson's Craftsmen' of Kilburn, Yorkshire, were engaged to make a suitable bible box of oak, to match some of the existing library furniture. This box, with its distinctive carved mouse trade-mark, is particularly appropriate as the home of 'Flora Norvegica' for it is carved from oak, brought down in the great gale of October 1987 near Kew Gardens, South Kensington, and was an acorn in 1766, the year of the publication of Part I of Gunnerus' book.

The oak-tree, SOMMEREIK, *Quercus robur,* has been my favourite tree for seventy years, ever since, as a tiny child, I had to recite of its hospitable, strong branches. Bishop Gunnerus had been renowned for his generous hospitality, several members of his family had appealed to him and never in vain, and he had housed and nurtured the Royal Society's Library.

The copy of 'Flora Norvegica' was in need of restoration, some pages were loose, the corners of the cover were damaged and the spine somewhat torn. The University of Birmingham book restoration department was prepared to undertake the task.

By the time the book was returned to its oak home, I was an Honorary Fellow of the college and had a suite of rooms in the Hall where the college had commenced residential land-based courses in 1948. I was in residence on the exciting afternoon of Gunnerus' homecoming. The restorer was sad to part with the fascinating book and told me some of the details revealed.

Cotswold Court, Warwickshire College. Flowering Rowan (Mountain Ash) seen on upper right.

The pages had to be completely, gently washed three times, proving that the handmade paper was made from 'rags' and not wood pulp. They were a joy to handle, so soft yet firm. Rodney Hobbs could not tell me the consistency of the 'rags' – a term covering a variety of materials in the twentieth century but only sparsely available in the eighteenth. When we consider the materials available then in Scandinavia it is likely to be old linen from local flax (*Linum usitatissimum*) and precious, well used, imported cotton.

The Warwickshire Window in the Learning Resource Centre at Moreton Morrell. Design inspired by the Italiante Garden at Moreton Hall - the view from the previous Looi Horticultural Library.

Flora Norvegica

Dipsacus fullonum
N. Kardeborre
E. Teasel

The open book - detailed example next page.

The rags were 'teased', using metal carders (combs) or dried teasel flower heads, *Dipsacus fullonum* (Norwegian – KARDEBORRE) into shreds and ground into a pulp with a pestle, a club shaped beater of stone, in a stone bath. The required size of mold, consisting of a fine mesh wire held in a wooden frame, was lowered into the pulp slightly below surface. It was then gently raised, strained, and covered with a sheet of felt the same size as the frame and inverted. The mold was removed and the paper pulp pressed between the felt, to absorb excess moisture and then left to air dry. Sometimes the surface was smoothed by polishing with a metal tool, stone or passed between wooden rollers. If one holds the treated pages of 'Flora Norvegica' to the light today one can still see the faint, faint lines where the wires of the mesh had been nearly three hundred years ago.

The restorers had found, inside the book, the original leather title from the spine of the book. The gold lettering could not be restored but it had been incorporated in the new leather spine and the leather cornered, marble boards had been retained. It could be presumed the book had been rebound at least twice before for the original copy would be bound in leather, which disintegrates with time and use.

Four pages were printed at a time and it is possible to see the markings that provide this information. The pages of the two books, Part I of 1766 and Part II published posthumously in 1776, now had the inner edges strengthened with Japanese fine tissue and the sections were hand-sewn. The inner spine had completely disintegrated and most pages were loose.

CARUM carvi. Detailed example of 'Flora Norvegica'.

CARUM CARVI
N. Karve
E. Caraway

Carum carvi (dried) was a major export from Trondheim in the seventeenth and eighteenth centuries. It was used as a preservative in flour-based breads and confectionery, and flavouring in cheese.

Bishop Johan Ernst Gunnerus' 'Flora Norvegica' is now safely contained, under ideal, monitored conditions.

Tomorrow some student is going to ask,

"Why should we treasure this old book in Warwickshire?"

I hope this account will assist the search and each reader find some 'seed' that will bear 'fruit'.

12

POSTSCRIPT
2002

The season of spring, in the temperate zones of the world, annually demonstrates the resurgence of new life. Colour has gradually faded from plants and trees and the once vivid, golden – green landscape has merged, through distant mauve, into a leafless etching of black, grey and brown. A thick, sheepskin-like blanket of snow will stretch over the coldest parts for some months keeping roots and seeds below ground warm. Progressively, as the sun gains strength, the snow will release moisture to permeate the ground and stimulate germination. Then little by little spears of green stretch out to the light from their burial ground, just as the chicken emerges from the apparently dead egg. Life above ground is resumed.

In equatorial regions, because of the unchangeable length of daily sunlight, causing constant heat, and the frequent supply of rainfall, growth is rapid and all the stages of development are found intermingled on each tree throughout the year.

Postscript 2002

Spring beside the Fjord.
Betula pendula (Bjørk).

WINTER in the mountains lays a snow blanket over Hol Church.

1773 was not the end of Johan Ernst Gunnerus. I have endeavoured to trace his roots and growth in this simple account and realise how branches of thought and influence continue still, frequently in quiet, unrecognised ways, through people who have influenced our thinking and unconsciously directed our way and whose names have never been remembered. It was a chance remark that set me on this trail and unexpected encounters that enlightened some new aspect.

Gunnerus himself had completed the second part of 'Flora Norvegica' during 1772–73. He had written the introduction whilst in Copenhagen, dated 23rd April 1772, but it had not been published before his death. Niels Dorph Gunnerus, the orphaned nephew who had been taken into the Bishop's home as a student in 1763, saw to the publication of this volume. Niels was a member of the Royal Norwegian Society of Science and Letters and continued his uncle's interest in natural science. He was a civil servant holding positions in Trøndelag and later in Bergen.

I was curious when I came across the name of Gunnerus' amanuensis, Jens Finne Borchgrevink, and thought perhaps his mother's name had been Finne. As President of the Singapore Gardening Society I was introduced to a Norwegian couple temporarily resident in the Republic and keenly interested in horticulture. Over twenty-five years our friendship has deepened, although they are now 'retired' in Oslo. They are called Finne. Wouldn't it be wonderful if they were descended from the same family? No, they were not but the question led to other interesting contacts and they have been my most interested and capable amanuenses in this work.

The search made in the church archives with regard to the Borchgrevinks made me realise what an astonishing town Røros was. I had briefly visited the town twice before but it was only on going in the steps of Gunnerus, to find his ex-student who became one of his special assistants, that I became truly aware of this fascinating area.

I was intrigued by the simple splendour of the church at the summit of Church Street *(Kjerkegata)*, fondly called, "Bergstadens Ziir", standing like the Temple of the Old Testament on Mount Zion. I wondered if the unusual word was a Norwegian translation of the Hebrew word, Zion, but was informed it is an archaic word implying 'pride and joy', a place, or person, of magnificence giving inward happiness.

Within the church, on a higher level than the pulpit, was the Baroque organ, which had been moved from the old church. The date of this instrument is uncertain but it was not new when it was built in the church of 1650. It is probable that the Borchgrevink family had a particular interest in church music for three ancestors of Leonard Christian Borchgrevink had been cathedral organists in the Danish Court and Norway. One of the first three organists in Røros was Johannes Brun who may have been the father of Magdalena Brun (born 1705). She was married to Leonard Christian Borchgrevink on 18th September 1727 and was subsequently the mother of Jens Finne Borchgrevink.

A new organ, built by Hollenbach, was installed in 1890 and the modern one now in use purchased in 1966.

AUTUMNAL changes in England. The coniferous, evergreen tree remains the same but within one week the beech has lost its leaves, with the help of the wind, and they now form a carpet. Beech Tree (deciduous in contrast to neighbouring Coniferous pine), Fagus sylvatica, N. Bøk.

Preparing for Christmas in Røros.

Postscript 2002

Svinvik Arboretum.
A large group of trees are "evergreen" for they have a different type of leaf which is not shed in winter. They hold their seeds in the petal-like folds of brown cones which open to release them when weather conditions are favourable. Their many "leaves" sticking out at right angles from every branch and off-shoot are short, dark green, smoothly rolled and sharply pointed like needles. The familiar "Christmas Tree" is a typical example.

Borchgrewinck – Brun – Finne Chapel.
March 2002.

Displayed on the walls, to the left and right of the pulpit, in full view of the congregation, are portraits of men who have played a prominent part in the development of Røros Copperworks and Bergstadens Ziir. The congregation refused one portrait! The miners had had to face Director Johannes Irgens and later his widow when alive but they would not do so after death. Their portrait was placed out of sight behind the altar.

Before the church was completed in 1784 the wealthy Catharina Borchgrevink built the Borchgrevink Chapel on the west side of the churchyard. The earthly remains of the Finne, Brun and Borchgrevink Families were then removed from the crypt beneath the old church. Her father had died in September 1772, shortly after Bishop Gunnerus' last visit.

When the new church was completed it became necessary to improve the main path to the church as worshippers were still taking the familiar way to the old church and continuing across the graveyard to Bergstadens Ziir. So the Borchgrevink Chapel now stands across the road from the main gate to the church.

This delightful picture, reproduced from the Røros Museum's publication on the occasion of the bicentenary of 'Bergstadens Ziir' in 1984, indicates so many interesting details. On the left is the Borchgrevink Chapel adjacent to the old school, on the right can be seen one of the two gateposts and the wall of the new church yard, whilst dipping away in the middle distance can be seen the miners houses. The dresses of the people clearly show the ages and wealth of the people and how the church was their social community centre.

Røros - gathering across 'Church Street'.

Catharina Borchgrevink built a new school in 1799 across the street from the church. The Copperworks had built the first school, like the church, but this was a charitable gift in which she took a personal interest until her death in December 1804, aged seventy-four.

Her youngest sister, Maria's second marriage was to Jens Finne in 1790, not to be confused with her step-grandfather, for there was no relationship whatever. Maria's husband was the son of Rasmus Finne (or Finde, both spellings pronounced the same), a priest in the diocese of Bergen.

Jens Finne Borchgrevink took up various positions in the church until his death in 1819 at Norderhov. As a boy and later student, he must have loved to go over the Linné Path in Røros, still signposted. Now his spirit must joyfully prompt others to visit the nearby Lake Aursunden where *Aster sibiricus*, SIBIRSTJERNE, has been found since 1975 in steadily increasing numbers, on a sandy bank close to the lake shore. How did the seed make that journey from the

WINTER AURLAND where Jens Finne Borchgrevink resided some years.

Postscript 2002

ASTER SIBIRICUS
N. Sibirstjerne

Looi Eng San Garden, Raffles Institution, Singapore.
Through the Bauhinia coccinea can be seen the armillary sphere, design based on a sundial I had first seen at North Cape.

Cypripedium calceolus
continues to be magnificently displayed, anywhere.

Kola Peninsular, Siberia, beyond the limits of the old Trondheim Diocese? How many thousands of years did it take to make that journey and how many centuries dormant? Perhaps it arrived by air, carried by birds.

In the same year as Jens Finne Borchgrevink died, another young, energetic man, Gunnerus-style, made a move at the other side of the world that was to change my life, and many others. In 1819 Stamford Raffles purchased the island of Singapura from the local agong (chief) of the neighbouring State of Johore, on behalf of the British East India Company. He had a short, stormy career but his perception, interests and wisdom shone through – with the aid of others. He was a keen, amateur natural scientist and an avid collector, especially whilst resident in Java. He was one of the founders of the London Zoological Gardens.

In Singapore his living memorial is in the prestigious school he founded in 1823, which retains the name of Raffles Institution. There is nothing 'institutional' about it today. 'Institution' is a word like 'Ziir' which has somewhat changed its connotation over the decades. Raffles Institution, Raffles Girls' School and Raffles Junior College provide the best opportunities for learning with residential and up-to-the-minute facilities in sport and education. The web of history continues to be woven, the ply sometimes twisting together to make stronger threads, but never completely broken in the Master Weaver's Hand.

The scientific skill and powers of observation of Oeder are still evident. The title 'Flora Danica' is well known but not in reference to the massive eighteen-volume work of botanical drawings commenced by Oeder but the exquisite Royal Copenhagen porcelain. The peacefulness, patience and concentration was enveloping as I toured the studio (factory does not seem the correct word for these works of art) recently in Copenhagen.

The method of manufacture of the fine porcelain is similar to such factories I have previously visited in the Midlands of England but here the decoration is completely applied by hand copied directly from one of the original specimen from the books 'Flora Danica'. Each flower on each piece of porcelain is a unique, perfect botanical specimen in every detail. Floral decorations were particularly popular in the nineteenth century but 'Flora Danica' stands alone for its botanical accuracy.

The authenticity goes further. Beneath each piece there is not only the familiar trade mark of the Royal Copenhagen Porcelain Factory, three parallel, wavy blue lines representing the three narrow Sounds which separate Denmark but also the Latin name of the specimen and the name of the person who first described it, beautifully inscribed. I feel to hold a sample of 'Flora Danica' I step back two hundred and fifty years and join Gunnerus and Oeder in Snåsa.

My most treasured piece is a wineglass bath manufactured to hold ice on an eighteenth century dining table, in which glasses could be dipped to chill or wash. It is decorated with sixteen different specimen and a large one covering the interior base, – in my possession, a wild rose, *Rosa villosa*. The name of each plant is inscribed on the base and here are two first described by Oeder (Oed), one by Vahl, a later professor

Royal Copenhagen Porcelain, Denmark.

'Flora Danica' wineglass bath.

Postscript 2002

Potentilla retusa Müller. From Flora Danica.

Rosenborg Palace, Copenhagen where the first full dinner service of 'Flora Danica' is displayed.

'I lift mine eyes to the hills.' Psalm 121

of botany in Copenhagen, many by Linnaeus (L) and one I was particularly pleased to have by Müller (Mull).

Otto Friderich Müller (1730–1784) wrote to Gunnerus, on receiving a copy of 'Flora Norvegica', to thank him and described him as 'the Bishop with the Bible in his right hand and one of Linnaeus' books, 'System Natura' in his left.' Linnaeus had also described Gunnerus as 'a great northern light'. What an enlightened revelation we have when we look with an inner light on nature around us and the Spirit within.

Sometimes grains of ideology and philosophy remain like seeds, *Aster sibiricus* or acorn, dormant, unnoticed by the human eye for centuries. Gunnerus and Schøning quietly stirred a patriotic spirit in Trondheim and through the work of the Royal Scientific Society hoped permission would be granted for the founding of a Norwegian University. Gunnerus had his hopes completely shattered on his visit to Copenhagen 1771–2.

You cannot kill a seed. When the time and conditions are right the outer shell will be broken and growth will ensue. One hundred and fifty years after Johan Ernst Gunnerus' death the grey granite, twin-spired building of the Norwegian School of Technology took shape. Always a renowned centre of science and technology (I remember the old railway engine in the forecourt in 1960) it grew from strength to strength.

Although the Act setting up the University of Trondheim was not passed until 1984, in 1974 the existing college library had merged with the oldest scientific library in Norway that of the Royal Scientific Society of Sciences and Letters, commenced in Gunnerus' home in 1768. An idea, which took almost two hundred years to germinate, is branching and blossoming. Now it is up to us!

Helen Keller, the American lady who became blind and deaf through illness as a tiny child, but overcame all obstacles through the constant patience of her teacher, said, 'The world is moved along, not only by the mighty shoves of its heroes, but by the aggregate of each tiny push of each honest worker.'

May we be stepping-stones, even if only small ones, to help some one on their way to see God's world around them, to handle the treasures with thoughtful understanding and to pass the knowledge on by thought, deed and word.

Therapeutic Geiranger

The Warwickshire College, Reception area, seen from the Alan Parton Teaching Block.

THIS PLANET

GOD in his love for us lent us this planet,
    Gave it a purpose in time and in space:
Small as a spark from the fire of creation,
    Cradle of life and the home of our race.

Thanks be to God for its bounty and beauty,
    Life that sustains us in body and mind:
Plenty for all, if we learn how to share it,
    Riches undreamed-of to fathom and find.

Long have our human wars ruined its harvest;
    Long has earth bowed to the terror of force;
Long have we wasted what others have need of,
    Poisoned the fountain of life at its source.

Each is the Lord's: it is ours to enjoy it,
    Ours, as his stewards, to farm and defend.
From its pollution, misuse, and destruction,
    Good Lord, deliver us, world without end!

F. Pratt Green (1903–2000)

LIST OF ILLUSTRATIONS

Ivar Mølsknes, Adresseavisen, the photography of bishop Gunnerus and the author in the library.
Andresen & Butenschøn page 17.
The Heaton Cooper Studio pages 18, 28, 33, 60, 61, 62, 64, 68, 69, 85, 98, 107, 110, 111, 112.
ExxonMobil pages 32, 52.
Universitetsforlaget page 44.
Trondheim Katedralskole page 49.
Nordenfjeldske Kunstindustrimuseum pages 51, 54, 116.
Kommuneforlaget page 56.
J. W. Cappelens Forlag page 58.
Gyldendal Norsk Forlag pages 65, 66, 88, 89, 99, 102, 126, 127, 134.
Per E. Fredriksen, Norwegian University of Science and Technology, the photography of Gunnerus' Microscope, page 78.
Rørosmuseet pages 84, 133.
The Linnéan Society of London page 93.
Stainer & Bell Ltd page 139.
The remaining photographs are by the author.

BIBLIOGRAPHY

Aase, M. 1998. Biskop i beste opplysningsånd. *Forskningspolitikk* 21, 12-13.

Aase, M. and M. Hård 1998. "Det norska Aten" Trondheim som lærdomsstad under 1700-tallets andra hälft. "Athens of the North" : Trondheim as an intellectual environment in the second half of the 18th century. *Lychnos* 1998, 37- 74.

Blunt, W. 1971. *The compleat naturalist : a life of Linnaeus*. London, Collins, 256 p.

Buckley, V. C. 1939. *Happy countries*. London, Hutchinson, 288 p.

Cooper, A. H. 1907. *The Norwegian fjords*. London, Adam and Charles Black, 178 p.

Daae, L. 1863. *Throndhjems Stifts geistlige Historie : fra Reformationen til 1814*. Throndhjem, J. Andersens Enke, 247 p.

Dahl, O. 1892-1911. *Biskop Gunnerus' virksomhed fornemmelig som botaniker: tilligemed en oversigt over botanikens tilstand i Danmark og Norge indtil hans død*. Trondhjem, 5 vol.

A Dictionary of eighteenth-century history. 2001. Edited by J. Black and R. Porter. London, Penguin, 880 p.

Engegård, G. 1973. Biskop Gunnerus og "Flora Norvegica". *Blyttia* 31, 3- 15.

Gjærevoll, O. 1990. *Norges planteliv : fra Sørlandsskjærgård til Svalbardtundra*. Oslo, Aschehoug, 304 p.

Gunnerus, J. E. 1758. *Hans opvækkelige Hyrdebrev til det velærværdige, høj- og vellærde Præsteskab i Tronhjems Stift*. Trondhjem, Jens Christensen Winding, 40 p.

Gunnerus, J. E. 1761-1772. *Brevveksling 1761-1772. Johan Ernst Gunnerus og Carl von Linné; utgitt av Leiv Amundsen; med bistand av Rolf Nordhagen og Erling Sivertsen*. Trondheim, Universitetsforlaget, 205 p.

Gunnerus, J. E. 1776-1772. *Jo. Ern. Gunneri Flora Norvegica : observationibus praesertim oeconomicis, panosque norvegici locupletata*. Nidrosiae, Typis Vindingianis, 2 vol.

Gunnerus, J. E. 1997. *Hans opvækkelige Hyrde-brev*. Trondheim, Universitetsbiblioteket i Trondheim, 40 p. Facsimilia Bibliothecae Universitatis Nidrosiensis: 1.

Hansen, T. 1964. *Arabia felix : the Danish expedition of 1761-1767*. London, Collins, 381 p.

Hansen, T. 1976. *Det lykkelige Arabien : en dansk ekspedition 1761-67*. København, Gyldendal 380 p.

Hohnen, D. 2000. *Hamlet's castle and Shakespeare's Elsinore*. Copenhagen, Christian Ejlers. 116 p.

Hohnen, D. 2001. *Hamlet, Kronborg og Shakespeares Helsingør*. København, Christian Ejlers, 116 p.

Krovoll, A. 1985. *Catalogue of the J. E. Gunnerus herbarium*. Trondheim, University of Trondheim, The Museum, 171 p. Gunneria, 52.

Lauring, P. 1968. *A history of the Kingdom of Denmark*. Copenhagen, Høst & Søn, 274 p.

Linné, C. von 1971. *A tour in Lapland*. New York, Arno Press, 2 vol.

Linné, C. von. 1975. *Lapplands resa år 1732*. Edited by M. von Platen and C-O. von Sydow. Stockholm, 276 p.

De Maré, E. 1952. *Scandinavia: Sweden, Denmark, and Norway*. London, Batsford, 262 p.

Mossberg, B. 1995. *Gyldendals store nordiske flora*. Oslo, Gyldendal, 695 p.

Munck, T. 1990. *Seventeenth century Europe : state, conflict and the social order in Europe, 1598-1700*. Houndmills, Macmillan, 457 p.

Møller, J. 1982. *Borger i Holbergs København*. København, Sesam, 191 p.

Nelson, N. 1973. *Denmark*. London, Batsford, 195 p.

Norsk historisk leksikon : næringsliv, rettsvesen, administrasjon, mynt, mål og vekt, militære forhold, byggeskikk m. m. : 1500-1850. 1974. Edited by R. Fladby, S. Imsen and H. Winge. Oslo, Cappelen, 386 p.

Petersen, T. 1918. Utsigt over biskop Johan Ernst Gunnerus's liv og virksomhet med særlig henblik paa det Trondhjemske Videnskapsselskaps stiftelse. Pp. 7-61 in: *Johan Ernst Gunnerus 1718-26. Februar-1918. Mindeblade utgitt av det Kongelige Norske Videnskabers Selskab*. Trondheim.

Printz, H. 1918. Biskop J.E. Gunnerus som botaniker. Pp. 79-96 in: *Johan Ernst Gunnerus 1718-26. Februar-1918. Mindeblade utgitt av det Kongelige Norske Videnskabers Selskab*. Trondheim.

Renouf, J. 1996. *Alfred Heaton Cooper: painter of a landscape*. Grasmere, Red Bank Press, 157 p.

Skytte Christiansen, M. 1973. *Historien om Flora Danica*. København, Dansk Esso A/S, 67 p.

Stagg, F. N. 1952. *North Norway : a history*. London, Allen & Unwin, 205 p.

Stagg, F. N. 1953. *The heart of Norway : a history of the central provinces*. London, Allen & Unwin, 194 p.

Stagg, F. N. 1954. *West Norway and its fjords : a history of Bergen and its provinces*. London, Allen & Unwin, 245 p.

Stagg, F. N. 1956. *East Norway and its frontier : a history of Oslo and its uplands*. London, Allen & Unwin, 285 p.

Stearn, W. T. 1992. *Botanical Latin : history, grammar, syntax, terminology, and vocabulary*. Devon, David & Charles, 546 p.

Stearn, W. T. 1994. *Stearn's dictionary of plant names for gardeners : a handbook on the origin and meaning of the botanical names of some cultivated plants*. London, Cassell, 363 p.

Til opplysning : Universitetsbiblioteket i Trondheim 1768-1993. University Library of Trondheim 1768-1993. 1993. Edited by H. Nissen and M. Aase. Trondheim, Tapir, 287 p. Det Kongelige norske videnskabers selskab. Skrifter. 1993: 1.

Trondheim : one thousand years in the city of St. Olav. 1992. Edited by J. Sandnes et al. Trondheim, Strindheim trykkeris forlag, 70 p.

Trondheims historie : 997-1997. 1996-1997. Edited by J. Sandnes et al. Oslo, Universitetsforlaget, 6 vol.

Wagner, P. 1990. Icones Florae Danicae. Flora Danicas "Urteteignere" og Illumniationsskolen for Quindekiønnet". Pp. 93- 100 in: *Blomster fra sans og samling*. København, Rhodos, 221 p.